WITHDRAWN

The Mathematics of Hydrology and Water Resources

The Institute of Mathematics and its Applications Conference Series

Optimization, *edited by* R. Fletcher
Combinatorial Mathematics and its Applications, *edited by* D. J. A. Welsh
Large Sparse Sets of Linear Equations, *edited by* J. K. Reid
Numerical Methods for Unconstrained Optimization, *edited by* W. Murray
The Mathematics of Finite Elements and Applications, *edited by* J. R. Whiteman
Software for Numerical Mathematics, *edited by* D. J. Evans
The Mathematical Theory of the Dynamics of Biological Populations, *edited by* M. S. Bartlett and R. W. Hiorns
Recent Mathematical Developments in Control, *edited by* D. J. Bell
Numerical Methods for Constrained Optimization, *edited by* P. E. Gill and W. Murray
Computational Methods and Problems in Aeronautical Fluid Dynamics, *edited by* B. L. Hewitt, C. R. Illingworth, R. C. Lock, K. W. Mangler, J. H. McDonnell, Catherine Richards and F. Walkden
Optimization in Action, *edited by* L. C. W. Dixon
The Mathematics of Finite Elements and Applications II, *edited by* J. R. Whiteman
The State of the Art in Numerical Analysis, *edited by* D. A. H. Jacobs
Fisheries Mathematics, *edited by* J. H. Steele
Numerical Software—Needs and Availability, *edited by* D. A. H. Jacobs
Recent Theoretical Developments in Control, *edited by* M. J. Gregson
The Mathematics of Hydrology and Water Resources, *edited by* E. H. Lloyd, T. O'Donnell and J. C. Wilkinson

In Preparation

Mathematical Aspects of Marine Traffic, *edited by* S. H. Hollingdale
Stochastic Programming, *edited by* M. A. H. Dempster
Numerical Methods in Applied Fluid Dynamics, *edited by* B. Hunt
Mathematical Modelling of Turbulent Diffusion in the Environment, *edited by* C. Harris
Mathematical Methods in Computer Graphics and Design, *edited by* K. Brodlie
Recent Developments in Markov Decision Processes, *edited by* D. J. White, R. Hartley and L. C. Thomas
Analysis and Optimization of Stochastic Systems, *edited by* O. L. R. Jacobs, M. H. A. Davis, M. A. H. Dempster, C. J. Harris and P. C. Parks

The Mathematics of Hydrology and Water Resources

Based on the Proceedings of the Conference on Mathematics of Hydrology and Water Resources held at the University of Lancaster in July 1976, organized by the Institute of Mathematics and its Applications

Edited by

E. H. LLOYD

T. O'DONNELL

J. C. WILKINSON

University of Lancaster, Cartmel College, Bailrigg, Lancaster

1979

ACADEMIC PRESS

London New York Toronto Sydney San Francisco

A Subsidiary of Harcourt Brace Jovanovich, Publishers

ACADEMIC PRESS INC. (LONDON) LTD.
24/28 Oval Road,
London NW1

United States Edition published by
ACADEMIC PRESS INC.
111 Fifth Avenue
New York, New York 10003

Copyright © 1979 by
The Institute of Mathematics and its Applications

All Rights Reserved

No part of this book may be reproduced in any form by photostat, microfilm, or any other means, without written permission from the publishers

British Library Cataloguing in Publication Data
The mathematics of hydrology and water resources.
(Institute of Mathematics and its Applications.
Symposium proceedings series).
  1. Hydrology—Mathematics
  2. Hydraulic engineering—Mathematics
  I. Lloyd, E. H. II. O'Donnell, Terence
  III. Wilkinson, John Craven IV. Series
  551.4'8'0151   GB656.2.M34   78-75272

ISBN 0-12-453350-7

Printed in Great Britain by John Wright & Sons Ltd.,
The Stonebridge Press, Bristol.

CONTRIBUTORS

J. AMOROCHO, *Department of Civil Engineering, University of California, Davis, California 95616, USA.*

R.T. CLARKE, *Institute of Hydrology, Maclean Building, Crowmarsh, Gifford, Wallingford, Oxfordshire OX10 8BB.*

J.C.I. DOOGE, *Engineering School, University College, Upper Merrion Street, Dublin 2.*

E.H. LLOYD, *Department of Mathematics, University of Lancaster, Cartmel College, Bailrigg, Lancaster.*

K. ORD, *Department of Statistics, University of Warwick, Coventry.*

M. REES, *Department of Statistics, University of Warwick, Coventry.*

B. RYDZ, *Severn-Trent Water Authority*

J.C. WILKINSON, *Department of Operational Research, University of Lancaster, Gillow House, Bailrigg, Lancaster LA1 4YX.*

PREFACE

The mathematical concepts and techniques that have been developed for use in engineering have come into being as a result in many cases of a direct response by mathematically minded engineers to the needs of their discipline, and in others by applications-minded mathematicians who have adapted existing mathematical knowledge and invented new methods for specific engineering problems. The originality, applicability and general success of these efforts, and the extent of fruitful interaction between mathematicians and engineers, have been influenced by the interests, training and abilities of the contributors, and by their understanding of the nature of the relevant engineering problems and their potential for mathematical treatment. This is so well recognized in some traditional areas such as electrical and mechanical engineering that a mathematical treatment of topics from these fields often forms an integral part of the instruction of mathematical undergraduates in our universities.

In some of the less traditional areas however it is possible for professional mathematicians to be unaware of the nature of the problems involved. This is largely true of hydrology and water resources engineering. It is a field in which there has been a rapid expansion in the past three decades, not only in the number of its practitioners, but also - and, indeed, particularly - in the number and extent of the mathematical developments that have been initiated from inside the profession. This is not to say, of course, that there have been no significant contributions from non-hydrologists: interesting and important developments have come from mathematicians, probabilists and allied workers. However, the total involvement of the professional mathematical community has undoubtedly been smaller than has been the case in, for example, aeronautical engineering, and this must be attributed not merely to the relative degrees of priority which the western world attaches to hydrology and aeronautics but also to the relative ignorance amongst mathematicians of the attractiveness of hydrology as a field for their endeavours.

PREFACE

It seemed desirable to bring some of these features to the attention of the mathematical community in the hope of stimulating an interest in what might be for many a new and fertile field, and that this might perhaps be done by arranging a conference in which particular problems were discussed both by engineers (mathematical and otherwise) and by applied mathematicians of various kinds (including probabilists and statisticians). With the help of the Institute of Mathematics and its Applications and of the University of Lancaster they therefore arranged a three-day meeting with the advertised aim of encouraging "...communication and interaction between hydrologists and mathematicians in order to improve and extend the application of mathematical techniques to problems of hydrology and water resources directed mainly to those working in two areas: hydrology, together with the neighbouring fields of civil engineering, meteorology, etc.; and mathematics, together with probability and statistics."

The principal papers presented at the Conference are collected in this volume.

Emlyn Lloyd   April 1979

The Institute thanks the authors of the papers, the editors, E.H. Lloyd, T. O'Donnell and J.C. Wilkinson (University of Lancaster) and also Miss J. Fulkes and Mrs. S. Hockett

CONTENTS

DETERMINISTIC INPUT-OUTPUT MODELS   1
    by J.C.I. Dooge

OPERATIONAL ASPECTS OF WATER RESOURCE PROBLEMS   39
    by J.C. Wilkinson

STOCHASTIC STORAGE PROBLEMS: THE WATER MANAGEMENT   57
    BACKGROUND
    by B. Rydz

STOCHASTIC STORAGE PROBLEMS   73
    by E.H. Lloyd

SPATIALLY DISTRIBUTED VARIABLES IN HYDROLOGIC MODELLING   87
    by J. Amorocho

SPATIAL PROCESSES: RECENT DEVELOPMENTS WITH APPLICATIONS   95
    TO HYDROLOGY
    by K. Ord and M. Rees

MULTIVARIATE SYNTHETIC HYDROLOGY: A THEORETICAL   119
    VIEWPOINT
    by R.T. Clarke

DETERMINISTIC INPUT-OUTPUT MODELS

J.C.I. Dooge

(University College Dublin)

HYDROLOGIC PROCESSES AND SYSTEMS

Hydrology is concerned with the occurrence and movement of water in the hydrosphere - i.e., above, on and below the surface of the earth. The total amount of water in this hydrosphere remains constant but an appreciable amount of it is in the course of transformation from one form of water to another or of movement from one location of water storage to another. The hydrological cycle for the earth as a whole is driven by the energy available from solar radiation. Water balances can be drawn up not only for the earth as a whole but also for a continental region, for an individual catchment area (large or small), or for a small area of a catchment (e.g. for a single field or for a short length of roadway surface).

The hydrological cycle is usually depicted in a form similar to that shown in Fig. 1 taken from a well known reference work (Chow 1964),. An alternative representation which is more suitable for the present discussion is shown in Fig. 2, where the rectangles denote various forms of water storage: in the atmosphere, on the surface of the ground, in the unsaturated soil moisture zone, in the groundwater reservoir below the water table, or in the channel network draining the catchment. The arrows in the diagram denote the various hydrological processes responsible for the transfer of water from one form of storage to another.

Fig. 2 shows the relationship between the various forms of water storage and water movement. Thus the precipitable water (W) in the atmosphere may be transformed by precipitation (P) to water stored on the surface of the ground. In the reverse direction water may be transferred from the surface of the ground or of vegetation to the atmosphere by means

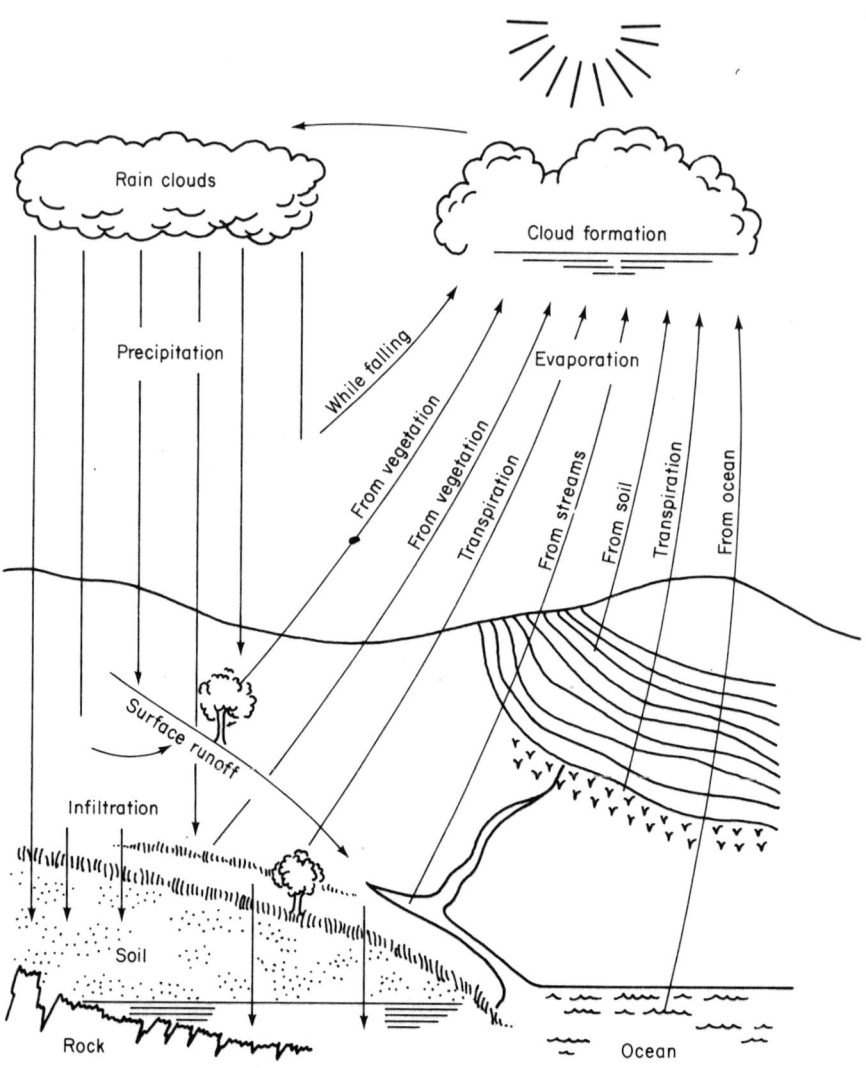

Fig. 1 Hydrologic Cycle (Chow)

of evaporation and transpiration (ET). Some of the water on the surface of the ground will infiltrate the soil through the surface of the soil (F) while some of it may find its way as overland flow (Q_0) into the channel network. During and following precipitation, soil moisture in the unsaturated subsurface zone is replenished by infiltration (F) through the surface. If the field moisture deficiency of the soil which had arisen since the previous precipitation period is substantially satisfied, there will be a recharge (G) to groundwater and also a certain amount of interflow (Q_i) or lateral flow through the soil which is intercepted by the channel network. The groundwater network storage is depleted by groundwater outflow (Q_g) which enters the channel network and supplies the streamflow during dry periods. During prolonged dry periods soil moisture may be replenished through capillary rise (C) from groundwater. Overland flow (Q_0), interflow (Q_i) and groundwater (Q_g) are all combined and modified in the channel network to form the runoff from the catchment. These various hydrologic processes are discussed in detail in textbooks and monographs on physical hydrology (e.g., Eagleson, 1969),.

It is not possible in practice to distinguish from an observed record of runoff the separate components of overland flow, interflow and groundwater flow discussed above. The most that can be done is to distinguish between a relatively rapid response of the catchment to a precipitation event and a second slower response. The quick response is often identified with surface runoff in the form of overland flow and interflow in the upper layers of the soil and the slower response with the passage of the water through the soil, both unsaturated and saturated. Accordingly, most models of catchment behaviour used in applied hydrology are elaborations of the simplified catchment model in Fig. 3 which considers the total response of the catchment to precipitation as including three subsystems: one representing direct storm response, the second representing groundwater storage and outflow, and the third representing the unsaturated soil moisture zone. The incoming precipitation (P) is divided into the effective precipitation (P_e) which produces the direct storm response, and the total infiltration (F). Since the amount of infiltration into the soil depends on the state of the soil there is a feedback (shown dotted in Fig. 3) from

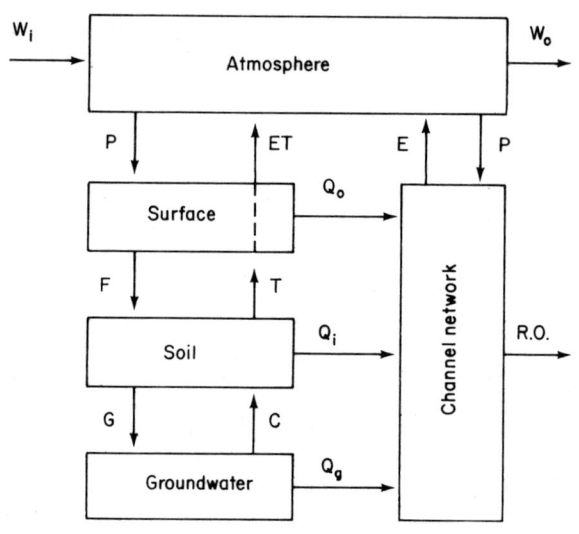

Fig. 2 Hydrologic Cycle for Catchments

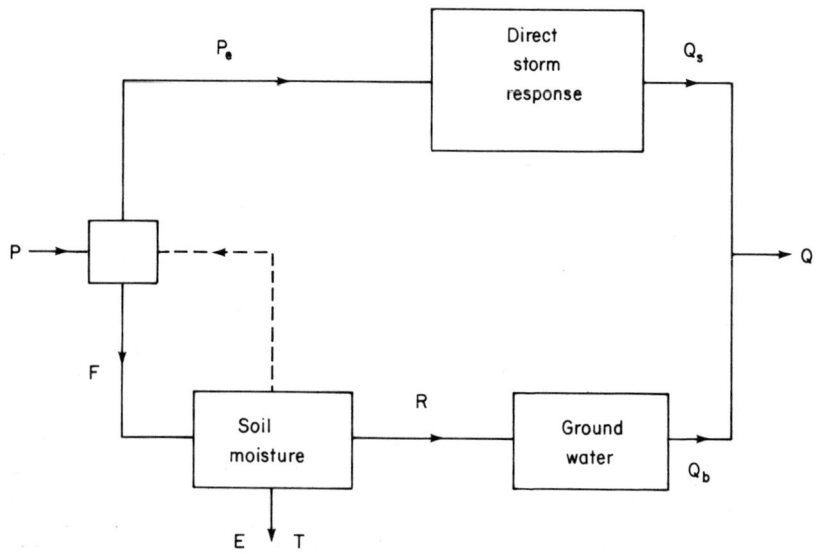

Fig. 3 Simplified Catchment Model

soil moisture to the divider which separates effective precipitation and infiltration. The soil moisture is considered as the only source of evaporation and transpiration and also as the source of recharge (R) to groundwater as a result of drainage of the unsaturated soil moisture zone.

The analysis of catchment behaviour can be approached from a number of points of view. Firstly, one could adopt what might be called a mathematical physics approach which would seek (a) to establish differential equations governing the physical phenomena involved, (b) to formulate the set of equations and boundary conditions for the particular catchment under study and (c) to attempt to solve the resulting problem for a given input of precipitation and other meteorological factors. Such an approach might use the St. Venant equations to describe both overland flow and the flow in the channel network, the Boussinesq equation for saturated flow in porous media to describe the groundwater flow and the Richards equation for unsaturated flow to describe infiltration, percolation and interflow in the soil above the water table. Since all of these equations are non-linear, their solution even for simple boundary conditions is a matter of difficulty and their solution for complex heterogeneous catchments presents great difficulties. A recent discussion of this approach with particular reference to subsurface flow has been given by Freeze (1972). A second and sharply contrasting approach is that known as black-box analysis in which our knowledge of the physics of the processes involved is ignored (at least initially) and instead an attempt is made to extract from past records of input-output events enough knowledge of the operation of the hydrologic system to serve as the basis for predicting the output from other specified inputs. The development of this approach and its relationship to the classical methods of applied hydrology has been reviewed (Dooge 1973).

Between these two extremes is the approach based on what are termed conceptual models. Although both of the approaches given above represent conceptual models in the general sense of that word, the term is usually limited in discussing hydrologic systems to conceptual models in the sense of simple arrangements of simple elements whose structure and parameter values are chosen to simulate the behaviour of the catchment under study. Rather than three isolated classes of model - black-box, conceptual models, equations of mathematical physics - there is in effect a complete spectrum of models ranging from pure black-box analysis which makes no physical assumptions to a highly complex analytical approach based on transport equations derived from continuum mechanics. This paper is concerned only with black-box analysis and with some simple conceptual models.

BLACK-BOX ANALYSIS OF HYDROLOGIC SYSTEMS

For the purpose of discussing hydrologic systems, a system can be defined (Dooge 1973) as:

> "any structure, device, scheme, or procedure,
> real or abstract
> that inter-relates in a given time reference,
> an input, cause or stimulus
> of matter, energy or information
> and
> an output, effect or response
> of information energy or matter"

In the pure black-box approach, attention is focused on the interrelationship between input and output without any reference to the nature of the physical processes involved in the transformation. This general model may be represented by

$$y(t) = H\,[\,x(t)\,] \qquad (1)$$

where $x(t)$ represents the input to the system, H represents the operation of the system and $y(t)$ represents the resulting output. There are three basic problems in black-box analysis. First there is the relatively straightforward problem of prediction where the input $x(t)$ and the nature of the operator H are both known and have to be combined to predict the output $y(t)$. Secondly there is the problem of system identification in which both the input $x(t)$ and the output $y(t)$ are known and it is required to find the nature of the system operation represented by H in equation (1). Finally there is the problem of signal detection in which the system operation H and the output $y(t)$ are both known and it is required to deduce the nature of the input $x(t)$. This paper is primarily concerned with the problem of system identification.

As in other approaches to process analysis, certain simplifying assumptions are made which facilitate the solution of the problem and the adequacy of these simple methods evaluated. Even if these simplified versions are inadequate, they frequently point the way towards the solution of the more complex problem. While x(t) and y(t) in equation (1) could be considered as vectors and thus able to represent multiple inputs and multiple outputs, it is more convenient in a first discussion to treat them as single variables. In systems terminology this is equivalent to assuming that the system is lumped, i.e., that the variables are not distributed in space but only in time. Making such a simplification corresponds to the decision to undertake the study of ordinary differential equations before studying partial differential equations.

The next obvious simplification is to limit our attention for the time being to linear systems. When this is done, it is easy to show that equation (1) becomes

$$y(t) = \int_{-\infty}^{\infty} h(t,\tau) \cdot x(t-\tau) \cdot d\tau \qquad (2)$$

where $h(t,\tau)$ is the impulse response function for the system, i.e., the output resulting from an input in the form of a Dirac delta-function occurring at a time τ. It is clear from equation (2) that, if $h(t,\tau)$ is known, the output can be predicted for any given input. Thus $h(t,\tau)$ completely characterises the linear system being studied.

If, in addition, the system is assumed to be time-invariant then the response at any time t will depend only on the time elapsed since the occurrence of the input and hence equation (2) can be written as:

$$y(t) = \int_{-\infty}^{\infty} h(t-\tau) \cdot x(\tau) \cdot d\tau \qquad (3)$$

in which the right hand side represents the well known operation of convolution. Thus, for a lumped, linear, time-invariant system the output can be obtained by convoluting the input and the impulse response of the system. To find the impulse response from a given input-output record, it is necessary to solve an integral equation of the form of equation (3).

In the case where the input and output are only available in sampled data form, a discrete form of the convolution equation is used to characterise the operation of lumped, linear, time-invariant systems. Where the input takes a histogram form, i.e., is constant over each interval of length D, and the output is sampled at the same interval, then the convolution equation takes the form:

$$y(sD) = \sum_{\sigma=-\infty}^{\infty} h_D(sD-\sigma D) \cdot X(\sigma D) \qquad (4a)$$

which can be written without ambiguity as

$$y(s) = \sum_{\sigma=-\infty}^{\infty} h_D(s-\sigma) \cdot X(\sigma) \qquad (4b)$$

In the above equation, $X(s)$ represents the values of the input volumes, $y(s)$ represents the output at discrete intervals, and h_D represents the ordinates at the same interval of the response of the system to a pulse of length D. It is the case of discrete data described by equation (4) that is examined in the present paper. If the input is an isolated one starting at t=0 and if the system is causal (i.e., the effect cannot precede the cause), then the limits of the summation become finite and equation (4b) can be written as

$$y(s) = \sum_{\sigma=0}^{s} h_D(s-\sigma) \cdot X(\sigma) \qquad (5a)$$

It is sometimes more convenient to write this equation of discrete convolution in the equivalent form

$$y(s) = \sum_{\sigma=0}^{\infty} X(s-\sigma) \, h_D(\sigma) \qquad (5b)$$

Equations (4) and (5) above represent a set of simultaneous linear equations which can be written in matrix form as:

$$y = Xh \qquad (6)$$

where y^T is the vector of known output ordinates $(y_0, y_1 \ldots \ldots \ldots y_p)$, h^T is the vector of unknown ordinates of the pulse response $(h_0, h_1 \ldots \ldots h_n)$, and X is the matrix formed from the vector of input values $(X_0, X_1 \ldots \ldots \ldots X_m)$ as follows

$$\begin{bmatrix} X_0 & 0 & . & . & . & 0 & 0 \\ X_1 & X_0 & . & . & . & 0 & 0 \\ . & . & . & . & . & . & . \\ . & . & . & . & . & . & . \\ X_m & X_{m-1} & . & . & . & . & . \\ . & . & . & . & . & . & . \\ 0 & 0 & X_m & . & . & X_1 & X_0 \\ . & . & . & . & . & X_2 & X_1 \\ . & . & . & . & . & . & . \\ . & . & . & . & . & X_m & X_{m-1} \\ 0 & 0 & . & . & . & 0 & X_m \end{bmatrix} \quad (7)$$

$(p+1, n+1)$

The problem of system identification for a lumped, linear, time-invariant system is the problem of solving the above set of simultaneous linear equations for the unknown values of h.

In applied hydrology, the solution of this identification is complicated by the fact that the input and output data are subject to measurement errors and that there may be further errors because of the approximations involved in the assumptions of linearity and time-invariance. The presence of such errors in any inverse problem becomes of serious significance when the mathematical system being inverted is an ill-conditioned one. Unfortunately, in the case of the heavily damped systems encountered in hydrology the output is a much smoother function than the input and the mathematical inversion of the process is accordingly inherently unstable. Consequently, the solution of the problem of identifying the system (i.e., finding the impulse response or pulse response) is far from trivial. Any proposed method of solution must be proved to be reliable in the sense that it must be robust in the presence of errors in the input and output data.

PLAN OF NUMERICAL EXPERIMENTATION

When a record of input x(t) and output y(t) is available for a hydrologic system, it is always possible to assume that they system acts in a linear time-invariant fashion and to use some method of deconvolution to determine the pulse response $h_D(s)$ defined in accordance with equation (5). Unless the method of deconvolution used is grossly unsuitable or inaccurate, the reconstructed output obtained by convoluting the input and the estimated pulse response will approximate closely to the recorded output. Unfortunately, however, the degree of correspondence between the predicted and recorded output may be a poor indicator of the correspondence of the estimated pulse response to the true pulse response or of the ability of the estimated pulse response to predict the output from an input dissimilar to that from which the pulse response was derived. In the case of a hydrologic system which is truly linear and time-invariant, the fundamental problem is to determine the true signal of the actual pulse response in the presence of the noise created by errors in the input and output data. In the case of hydrologic systems for which linearity and time-invariance are only approximations, the fundamental problem is to obtain as good estimate as possible of the optimum linearised representation of the non-linear system.

Hydrologists working on the development of methods for linear system identification are aware of the mathematical significance of the values of the determinants and eigenvalues and condition numbers of the matrices they are seeking to invert. However, no comprehensive application of these concepts to hydrologic problems has appeared in the hydrological literature or in sources usually read by hydrologists. This means that hydrologists are ignorant as to whether the special form of the X-matrix given by equation (7) or of the Toeplitz-type matrix given by $X^T X$ would result in properties that might suggest certain methods of the deconvolution of equation (5) being more suitable than others.

In the absence of such theoretical information, research hydrologists have had recourse to numerical experimentation. In this connection a three-step strategy has been followed in developing and evaluating methods of system identification. In the first step the validity of a proposed identification method is verified by applying it to a synthetic set of input-output data generated by choosing a specific system response

Deterministic Input-Output Models

and convoluting this with a chosen input in order to generate the corresponding synthetic output. The impulse response or pulse response may then be estimated by applying the proposed identification method to the synthetic input and output data and comparing the derived system response to the known system response used in the generation of the data. If the variation in system response is appreciable, this indicates either some defect in the proposed method, some error in applying it or an undue amount of roundoff error in either the generation of the data or the use of the identification method. The second step consists of the verification of the robustness of a method of system identification (which has been found to be valid in the first step) by examining the effect on the results of errors in the input and output. This is done by adding to "error-free" input and output data error of a known type and magnitude and testing the ability of the method to derive the true unit hydrograph in the presence of such error. Only after the validity and robustness of the method of system identification have been verified as described above is the method applied to actual field data.

The first comprehensive study on the effect of errors on unit hydrograph derivation is that of Laurenson and O'Donnell (1969). In that study the authors assume the impulse response of the system to be

$$h(t) = \left[\left(\frac{1}{t + T} - \frac{1}{20 + T} \right) \exp\left(\frac{-20}{A \cdot t + T} \right) \right]^R \quad (8)$$

for all values of t between 0 and 20 and to be zero outside those limits.

The model represented by equation (8) is a three parameter one and could be used to generate a wide variety of shapes of response but in their study Laurenson and O'Donnell experimented with only two sets of the values of the parameters T, A and R. In this original study three shapes of rainfall, all of histogram form, were used. The combination of three alternative input patterns with two alternative shapes of input response gave rise to six sets of synthetic output data. Three methods of black-box analysis (least squares, harmonic analysis and Meixner analysis) and one conceptual model (a cascade of equal linear reservoirs) were applied to the synthetic data and their performance with error-free data tested.

The synthetic error-free data were then contaminated by systematic error of a type and magnitude likely to occur in hydrologic measurements. In all, six cases of systematic error were studied and their comparative effect on the different methods of system identification studied (Laurenson and O'Donnell 1969).

The pioneering study of Laurenson and O'Donnell was extended by Garvey (1972) who tested nine methods of black-box analysis and three conceptual models for their stability in the presence of six types of random error as well as the six types of systematic error previously studied. Garvey also investigated the effect of the shape of the unit hydrograph on fitting of conceptual models by using seven sets of parameters in the unit hydrograph equation given by equation (8). He also studied the effect of three different levels (5%, 10%, 15%) of error in the data on the mean error in the unit hydrograph (Garvey 1972).

Bruen (1976) is currently extending Garvey's investigation and computer program further by increasing the number of inputs studied from three to six, the number of methods of black-box analysis from nine to fifteen, the number of conceptual models from three to twenty-five, the number of types of random error from six to twelve and the number of levels of error studied from three to six. He is also extending the study to compute the mean and variance of a large number of realisations for each case of random error rather than individual realisations of the random process as was done by Garvey (1972). Another point being studied in Bruen's project of numerical experimentation is the effect of "filtering" either the input-output data (pre-filtering) or the estimated pulse response (post-filtering). A filter in this sense is an operation which removes or reduces an unwanted characteristic in the record. Thus the truncation of the Fourier series representation of a function or of a data series removes contributions from frequencies above the cut-off frequency and is a numerical frequency filter equivalent to an ideal low-pass filter with the same cut-off frequency. In the present discussion only post-filtering will be referred to. The most important post-filtering operations are (1) smoothing the pulse response by a moving average filter in the time domain or by a cut-off filter (e.g., a Blackman filter) in the frequency domain (filter S), (2) maintaining non-negativity by setting all negative ordinates equal to zero (filter N) and (3) imposing a mass continuity condition by normalising the

Deterministic Input-Output Models

sum of the ordinates of the pulse response (filter A).

In the remainder of this paper, the instability of the process of deconvolution will be illustrated on the basis of Bruen's results for a single input shape, a single form of unit hydrograph, a single type and level of error, a single package of post-filters and a limited number of black-box methods of analysis and conceptual models. The input chosen was one of those used by Laurenson and O'Donnell (1969) for which the input $x(t)$ in the interval from $t=i$ to $t=i+1$ is defined by

$$x(t) = 0 \cdot 01 + 0 \cdot 02 \text{ i} \quad 0 \leq i \leq 9 \tag{9}$$

which gives for the period from $t=0$ to $t=10$ a histogram form of linear late-peaking input. The unit hydrograph was that obtained from equation (8) by specifying the values $T=2 \cdot 5$, $A=3 \cdot 0$ and $R=2 \cdot 5$. The chosen input pattern and specified unit hydrograph were convoluted in order to produce a synthetic output and this output was then contaminated by the addition of Gaussian white noise at value of $\sigma = 0 \cdot 15 \bar{y}$ which is equivalent to a mean absolute error of about 10% of the average output. In each case 100 realisations of this error process were computed and the estimated unit hydrograph found for each realisation and compared with the unit hydrograph originally specified. The post-filters used in the numerical experiments were (a) the condition of non-negativity of ordinates (N); (b) smoothing of the derived unit hydrograph by a non-recursive linear 3-point symmetrical filter with values of $0 \cdot 40476$, $0 \cdot 19048$, $0 \cdot 40476$, and (c) normalisation of the area of the unit hydrograph to unity (A). For convenience the methods studied have been grouped as follows: (a) methods involving direct matrix inversion, (b) methods involving optimisation either unconstrained or constrained and (c) transform methods of solution. In each case the methods are described in outline, the mean r.m.s. error in both the reconstituted output and the estimated pulse are given and figures shown of particular realisations at the mean level of error.

RESULTS FOR SOLUTION BY DIRECT MATRIX INVERSION

One obvious method of solution of the problem of system identification, i.e., of the deconvolution of equation (6), is to solve for the unknown values of the pulse response vector h by matrix inversion. The number of equations in the system is determined by the number of ordinates in the output vector $(y_0, y_1 \ldots\ldots\ldots y_p)$. If the number of unknown ordinates of the system response is taken equal to the number of output ordinates then the matrix X is square and if it is non-singular can be inverted. However, consideration of the definition of the pulse response indicates that the number of ordinates in the output $(y_0, y_1 \ldots\ldots y_p)$, the number in the input $(x_0, x_1 \ldots\ldots x_m)$, and the number of ordinates in the pulse response $(h_0, h_1 \ldots\ldots h_n)$ are connected by the equation:

$$p = m + n. \quad (10)$$

Accordingly the number of ordinates in the pulse response (n+1) will be less than the number of ordinates in the output and we can write

$$h_i = 0 \text{ for } i > n + 1. \quad (11)$$

The elimination of these values of h_i involves the elimination of the corresponding columns of tne input vector X, thus reducing it from a (p+1, p+1) matrix to a (p+1, n+1) matrix.

The reduced matrix X obtained by making the assumption of equation (10) can be solved by direct matrix inversion by choosing any n+1 of the rows and inverting the resulting square matrix. It can be seen from equation (7) that, if the first (n+1) rows are chosen, then the matrix to be inverted will be lower triangular and can be solved directly by forward substitution. Similarly, if the last (n+1) equations are taken then the matrix to be inverted is upper triangular and the problem can be solved by backward substitution. If any other set of equations were taken, then some more complex algorithm for matrix inversion must be used.

If forward substitution is used for the case of error free data in order to find the estimated pulse response for the input and output data described in the last section then the r.m.s. error in the derived unit hydrograph as a proportion of the mean response ordinate has a mean value of 2×10^{-13}. This error is due to round-off in the computation

Deterministic Input-Output Models 15

process. If, however, the output is subject to random error with standard deviation equal to 15% of the mean output (i.e., a mean error of 10%) then the error in the derived unit hydrograph has a mean value equal to 4.7×10^3 times the mean ordinate, i.e., there is complete numerical explosion. If on the other hand the last (n+1) equations are taken and backward substitution used, the mean r.m.s. error in the unit hydrograph for error free data is 0·556 times the mean. If the shape of input used had been one of gradually decreasing input rather than gradually increasing input as defined by equation (9), then the position would have been reversed and forward substitution would have produced a superior result to backward substitution. Since, to be reliable in practice, a method is required to handle all types of input, it is clear that neither forward substitution nor backward substitution is adequate for the task.

The strong effect of the shape of input on the stability of the inversion process has been known to hydrologists for a long time. Such a result is not unexpected if one takes the value of the determinant as a measure of the stability of the solution of the inversion problem. The determinant of the matrix for forward substitution is $(x_0)^{n+1}$ and for backward substitution is $(x_m)^{n+1}$ when the input ordinates $(x_0 \ldots\ldots x_m)$ have been normalised for unit area. In the case of forward substitution the relatively small value of $x_0 = 0.01$ produces a determinant value of the order of 10^{-40} and in the case of backward substitution the relatively high value of $x_m = 0.19$ produces a determinant value of 4×10^{-15}.

The problem of sensitivity to input shape can be overcome to some degree by adapting a procedure proposed many years ago by an American hydrologist and to use the (n+1) equations starting with the first equation which contains the maximum input ordinate (Collins (1939)). An examination of equation (7) reveals that this ensures that the diagonal elements of the matrix to be inverted are greater than the off-diagonal elements. Even though the solutions due to the Collins method (and in the present case for backward substitution) are far superior to those of forward substitution, they still leave much to be desired. Fig. 4 shows the solution by the Collins method where the only constraint is that the length of the pulse response was taken at its correct value of 20 units, for a case where the standard deviation of the error on the output was 15% of the average output. The estimated pulse response in that figure shows negative ordinates and an oscillatory shape which does not occur in the original pulse responses which is also shown in Fig. 4. In addition it can be shown that the area of the pulse response differs from unity thus violating the continuity condition. It is possible to improve the estimated pulse response by imposing the constraints of non-negativity of ordinates, smooth shape and unit area on the derived pulse response as post-inversion filtering operations. When this is done for the case of the Collins method the mean r.m.s. error in the pulse response is reduced from 0·56 times the mean to 0·30 times the mean but the resulting hydrograph shown in Fig. 5 is still not entirely satisfactory.

These cases among others are given in Table I, which shows the mean r.m.s. error in the reconstituted output and in the estimated pulse response for the various methods. It will be noted from Table I that (a) all the methods perform well for error-free data; (b) for the input pattern studied, backward substitution and the Collins method are far more stable in the presence of data error than forward substitution; (c) post-filtering of the estimated pulse response improves the estimate but reduces the accuracy of reconstitution of the observed output.

RESULTS FOR METHODS BASED ON OPTIMISATION

The obvious starting point for any discussion of an optimisation approach to the problem of system identification is the method of least squares. While the methods of solution described in the last section seek to satisfy exactly (n+1) of the available (p+1) equations, the least squares method seeks a solution that will be a best fit to all (p+1) equations in the sense that the sum of the squares of the differences between the predicted and measured outputs will be minimised. This will certainly give a smoother approximation to the whole range of output but what we are concerned with in this discussion is whether it will give a better approximation to the system response. It can be shown that the least squares solution is obtained by the solution of the equation represented by

$$X^T y = (X^T X) \cdot h \qquad (12)$$

through the inversion of the square matrix $(X^T X)$. Whether

TABLE I

Results based on direct matrix inversion

Method of Identification	Post-filter	Error in observed output	Mean r.m.s. error of 100 realisations	
			Reconstructed output	estimated response
Forward substitution	none	error-free	3×10^{-14}	2×10^{-13}
Backward substitution	none	error-free	2×10^{-16}	5×10^{-16}
Collins method	none	error-free	2×10^{-16}	5×10^{-16}
Forward substitution	none	10% error	$9 \cdot 2 \times 10^2$	$4 \cdot 7 \times 10^3$
Backward substitution	none	10% error	0·092	0·556
Collins method	none	10% error	0·095	0·557
Collins method	NSA	10% error	0·193	0·302

The variance of these results for the case of forward substitution is of the order of 10^7. In all the other cases the variance is in the range 10^{-2} to 10^{-3}.

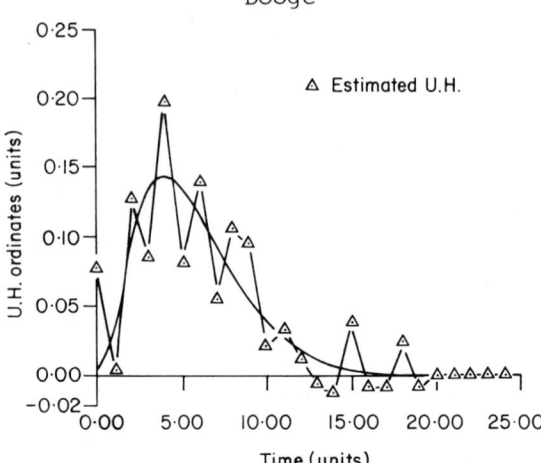

Fig. 4 Collins Method (NO FILTERS)

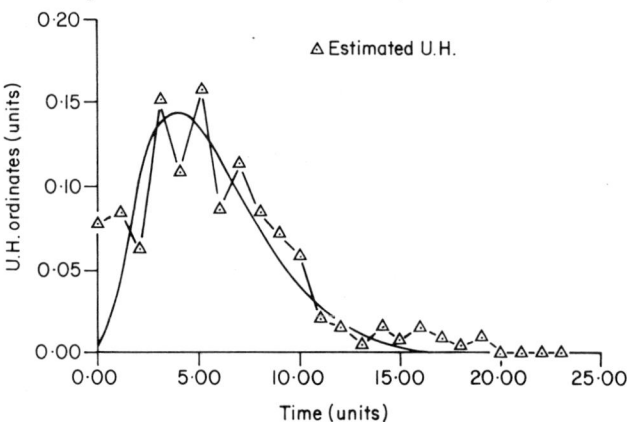

Fig. 5 Collins Method (POST-FILTER)

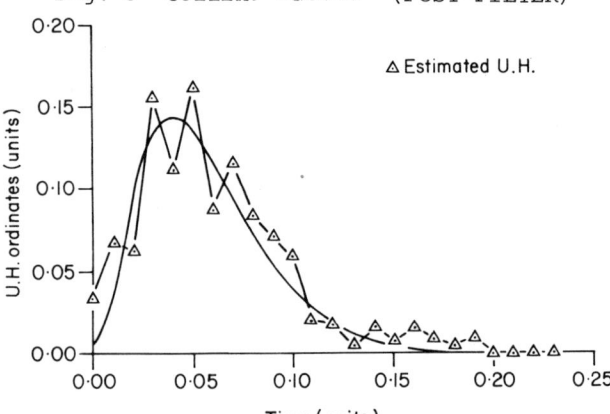

Fig. 6 Least Squares (POST-FILTER)

this will give an improved method of solution to the system identification problem depends on whether the latter matrix is better conditioned than the original matrix from which it was derived. The least squares method was applied to unit hydrograph derivation by Snyder (1955) and by Body (1959) and later improved by Newton and Vinyard (1967).

For the input-output event which is used on a basis of comparison in the present discussion, the least squares method gives results that are slightly but not appreciably better than those obtained by backward substitution or by the Collins method. Examining these results, it is not surprising to find that the square root of the determinant for the matrix X^TX) is 7.5×10^{-15} compared with 3.8×10^{-15} for the case of the Collins method. When the derived system response is adjusted by reducing all negative ordinates to zero, smoothing the resulting curve and normalising the area the result is that shown on Fig. 6 which is seen to be similar to that for the Collins method shown on Fig. 5.

In recent years, further developments based on the least squares method have been used in the identification of hydrologic systems. These involve the incorporation of the constraints applied as post-filtering operations in the methods described above into the inversion procedure itself. The basic procedure in these optimisation methods is shown in Fig. 7. The method of regularisation was applied to hydrology by Kuchment (1967) and that of quadratic programming by Natale and Todini (1973). The relationship between the three approaches is shown on Fig. 7.

The constraints of smoothing of various types and of unit area can be expressed in the form

$$Ch = b \qquad (13a)$$

where h is the vector of unknown ordinates of the pulse response. As long as the constraints are linear, the separate constraints can be added to give a single over-all constraint of the type given by equation (13). For the smoothing used in the present experiments we have

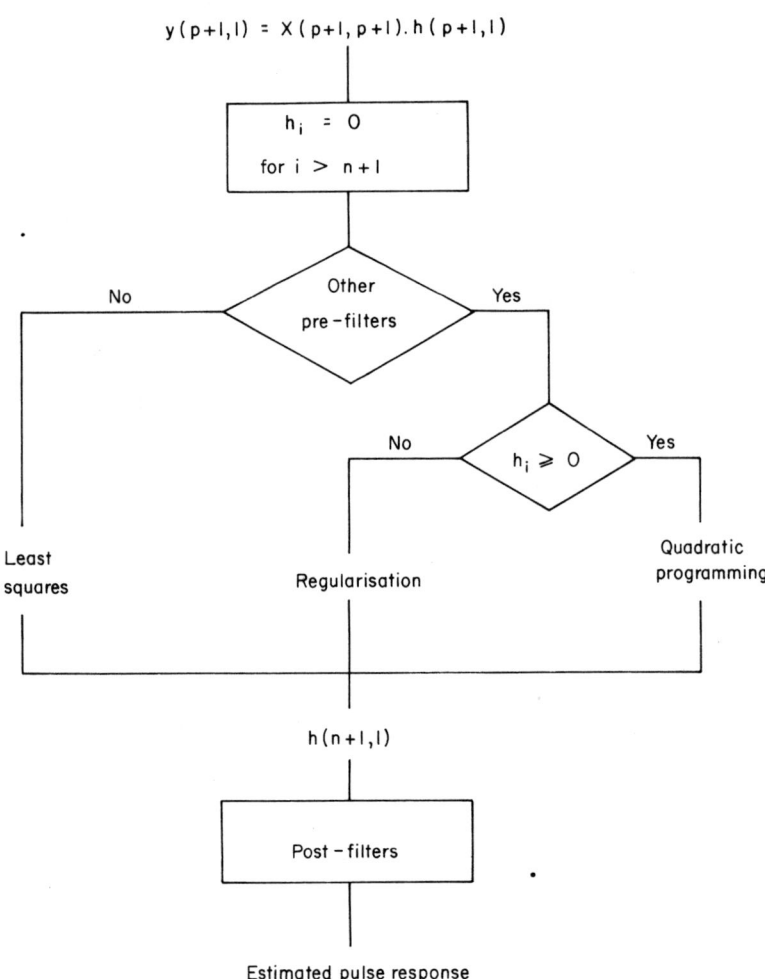

Fig. 7 Optimisation Methods

$$C = \begin{bmatrix} W_0 & W_1 & 0 & 0 & 0 & 0 \\ W_{-1} & W_0 & W_1 & 0 & 0 & 0 \\ 0 & W_{-1} & W_0 & W_1 & 0 & 0 \\ \cdot & \cdot & \cdot & \cdot & \cdot & \cdot \\ \cdot & \cdot & \cdot & \cdot & \cdot & \cdot \\ 0 & 0 & 0 & 0 & W_0 & W_1 \\ 0 & 0 & 0 & 0 & W_{-1} & W_0 \end{bmatrix} \qquad (13b)$$

With W_0 = 0·19048 and $W_{-1} = W_1$ = 0·40476 and b is a zero vector of the required length. For the unit area constraint every element of c is unity and every element of b is also unity. The solution which is optimal in the least squares sense for the simultaneous approximation of equation (6) and constraint (13) is that which minimises the objective function given by

$$(y - Xh)^T(y - Xh) + \gamma (b - Ch)^T(b - Ch) \qquad (14)$$

where γ is a weighting factor which reflects the relative weight given to the desire to satisfy the original set of equations given by (6) or the constraint given by (12). The general solution to equation (13) is given by

$$h_{opt} = (X^TX + \gamma \cdot C^TC)^{-1}(X^Ty + \gamma C^Tb). \qquad (15)$$

If the value of γ aproaches zero, then we have the original problem of finding the least squares solution to equation (6), without reference to the constraint. If on the other hand, γ approaches infinity then we have the problem of finding the least squares solution for the optimal satisfaction of the constraint represented by equation (12) without reference to the original set of equations. In practice, the choice of γ is subjective and is taken as the smallest value which eliminates the undesirable features of the unconstrained least squares solution. Numerical filters can be used as before in an attempt to improve the estimate of the pulse response.

Fig. 8 shows the results for regularisation in which 3 point smoothing is implicit in the solution and post-filterings are applied. If the non-negativity constraint is made an integral part of the solution of the problem then we have a quadratic programming problem. In this case we seek a minimum of the augmented objective function defined by equation (14) subject to the requirement that the elements of the solution vector are non-negative. Fig. 9 shows the result of quadratic programming using an algorithm due to Wolfe (1959) where the conditions of non-negativity smoothness and area are all incorporated in the solution process.

Table II shows the results for the optimisation methods.

TRANSFORM METHODS OF SOLUTION

Both the approach based on direct matrix inversion and that based on optimisation discussed above seek a solution to the basic identification problem by determining the elements of the vector of unknown ordinates of the response function in the time domain. An alternative is to seek some other representation for the three discrete functions involved which may be more convenient for solution purposes. The fact that Laplace transform methods are widely used in the analysis of linear systems in electrical engineering suggests that the problem of discrete convolution should be tackled by means of the corresponding z-transform defined by:

$$F(z^{-1}) = \sum_{s=0}^{\infty} f(\cdot s) \cdot z^{-s} \qquad (16)$$

where the notation $F(z^{-1})$ is used instead of the more usual $F(z)$ in order to emphasise the fact that the transform is a polynomial in z^{-1}.

For this transformation (as for the Laplace transform) the process of convolution in the time domain corresponds to the operation of multiplication in the transform domain so that the relationship between input and output of linear time-invariant systems defined by equations (4), (5) or (6) above is given in the transform domain by

TABLE II

Results for optimisation methods

Error in observed output	Method of identification	Implicit constraints	Post-filters	Mean r.m.s. error of 100 realisations	
				reconstructed output	estimated response
Error-free	Least squares	none	none	1.3×10^{-15}	2.5×10^{-15}
10% error	Least squares	none	none	0.082	0.548
10% error	Least squares	none	NSA	0.191	0.295
10% error	Regularisation	S	none	0.227	0.316
10% error	Regularisation	S	NSA	0.168	0.280
10% error	Quadratic programming	N	none	0.087	0.474
10% error	Quadratic programming	NSA	none	0.102	0.294

The variances of the results in Table II lie in the range 10^{-2} to 10^{-3}, with the exception of the error-free case. It will be noted that the more complex methods of regularisation and quadratic programming improve the reconstitution of the output but do not improve appreciably the estimation of the underlying pulse response given by the unconstrained method of least squares.

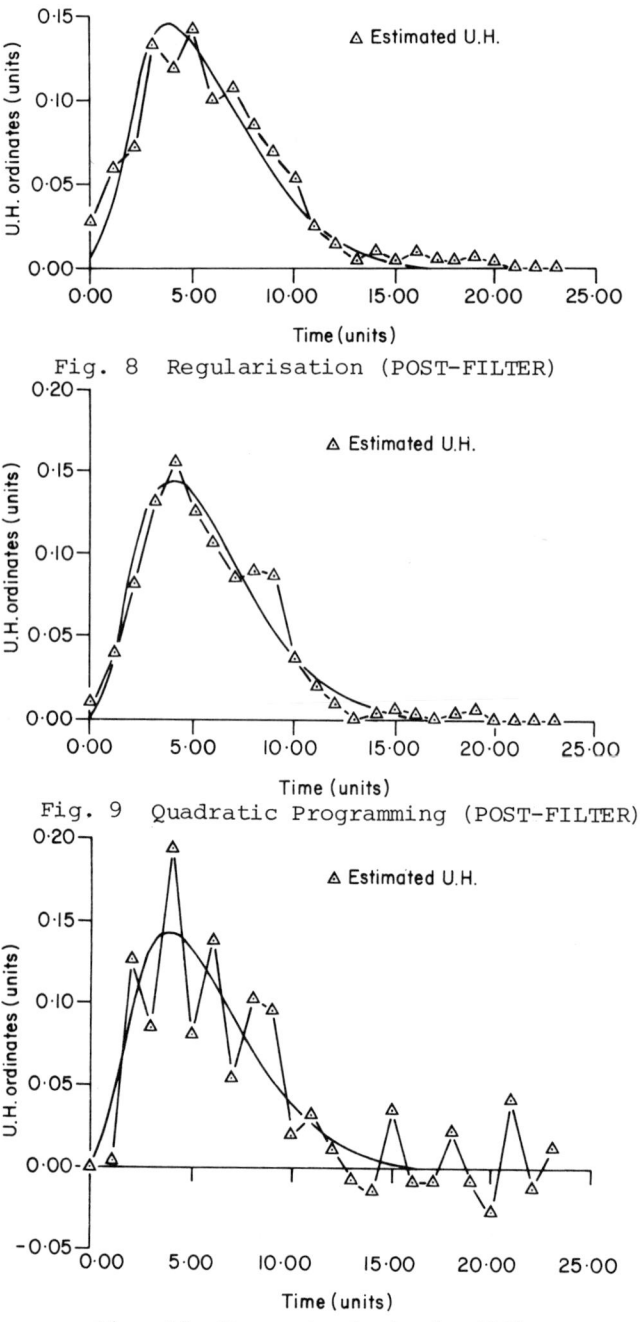

Fig. 8 Regularisation (POST-FILTER)

Fig. 9 Quadratic Programming (POST-FILTER)

Fig. 10 Harmonic Analysis (32)

$$Y(z^{-1}) = X(z^{-1}) \cdot H(z^{-1}) \qquad (17)$$

where $H(z^{-1})$ is the z-transform of the pulse response and is often termed the discrete system function.

An inspection of equation (15) suggests that we write

$$H(z^{-1}) = \frac{Y(z^{-1})}{X(z^{-1})} \qquad (18)$$

and proceed by polynomial division in z^{-1} to find the coefficients of the various terms z^{-s} in $H(z^{-1})$, which are by definition the values of h(s). This straightforward method will give acceptable results for error-free data but will prove most unstable in the presence of errors in the data. In fact, it is quite easy to show that the solution by use of polynomial division of the z-transform is equivalent in every way to the solution by forward substitution already discussed under direct matrix inversion.

If, however, the z-transform polynomials are considered not in terms of their coefficients but of their roots, the prospects for a robust method of solution become somewhat brighter. This approach was applied by De Laine (1970) to the derivation of the unit hydrograph from a series of storms on the basis that the roots of the output polynomial which recur in every storm must belong to the system function rather than the input polynomial and these common roots can therefore be used to reconstitute the pulse response function in the time domain. This approach can also be used in the case of a single input-output event by matching the roots of the input polynomial with the closest roots in the output polynomial and reconstituting the pulse response of the system in the time domain from the remaining roots. In the presence of data errors, the polynomial roots will change in value and the matching of roots becomes a difficult and in some cases uncertain process. This method is likely to be more efficient when the bulk of the data error is in the input than in a case like the present example where only output error is involved.

Another approach which suggests itself is that of harmonic analysis. This method was applied to hydrologic systems by O'Donnell (1960). If the input, pulse response and output are all represented as finite Fourier series, then the complex Fourier coefficients of the output (C) of the input (c) and of the pulse response (γ) are connected by the simple relationship

$$C_k = c_k \cdot \gamma_k. \qquad (19)$$

The similarity between equations (17) and (19) is due to the fact that the coefficients in equation (19) for any particular value of k correspond to equation (17) with a value of z given by

$$z = \exp\left[\frac{2\pi ik}{n}\right] \qquad (20)$$

where n is the number of data points for the function concerned. The efficiency of this method of identification is distinctly improved if use is made of the Fast Fourier Transform algorithm (see Brigham 1974).

If a full set of harmonic coefficients are used in the analysis of the input-output event, then the estimated pulse response will contain all of the frequencies up to half the sampling frequency and will reproduce both the signal represented by the underlying true pulse response and the noise represented by data error. The result of such an analysis is shown on Fig. 10. If, however, the series is truncated then for the heavily damped systems encountered in hydrology, the expectation will be that the removal of the high frequency components will remove the greater part of the noise without undue impairment of the underlying signal. The result for the realisation used in Figs. 4 to 10 is shown on Fig. 11. From the latter figure it can be seen that in this case the truncation does have an appreciable and beneficial effect on the shape of the estimated pulse response. The use of post-filters does not improve and in some cases may make the estimate of the pulse response worse.

Some years ago I suggested that, for the case of heavily damped systems such as are encountered in hydrology, trigonometrical functions might be replaced by Laguerre functions which are closely related to the gamma distribution widely used in hydrology for fitting empirical unit hydrographs (Dooge 1965). In the case of discrete input and output data

Deterministic Input-Output Models

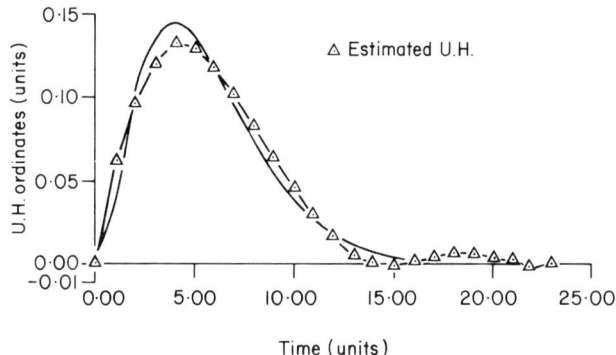

Fig. 11 Harmonic Analysis (9)

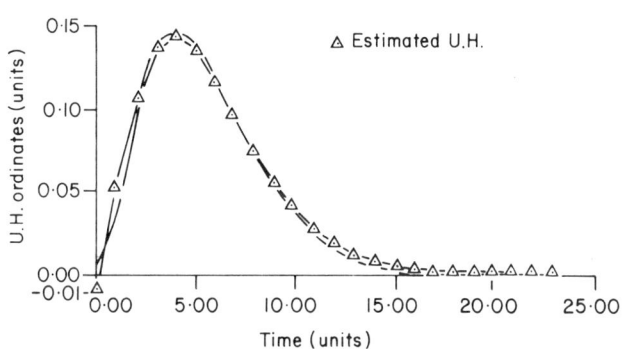

Fig. 12 Meixner Analysis (5)

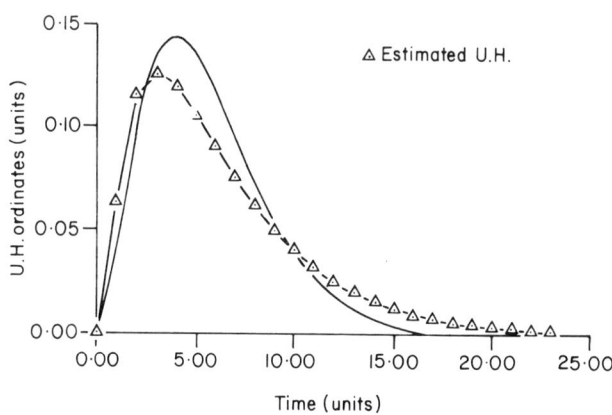

Fig. 13 N Equal Reservoirs

Laguerre functions are not suitable since (unlike the trignometrical functions) they are not orthogonal under summation as well as integration. For discrete data they must be replaced by Meixner functions defined by (Dooge, 1966).

$$\Phi_n(s) = (\frac{1}{2})^{\frac{s+n+1}{2}} \sum_{k=0}^{n} (-1)^k \binom{n}{k}\binom{s}{k} \quad (21)$$

which are orthonormal over the range $s = 0, 1, 2, \ldots \infty$. If the input, pulse response and output functions are all expanded as Meixner series, then the coefficients of the output (A), the coefficients of the input (a) and the coefficients of the pulse response (α) are connected by the linkage equation

$$A_p = \sum_{k=0}^{p} \sqrt{2}\, \alpha_k\, a_{p-k} - \sum_{k=0}^{p} a_{p-k-1}, \quad p = 0, 1, 2 \ldots \quad (22)$$

The matrix of coefficients formed from the input coefficients (a) and which multiplies the vector of unknown coefficients of the pulse response in equation (22) is seen to be of the same form as the input vector in equation (7). Any reliable method of matrix inversion or of optimisation can be used to solve equation (22). If the large number of Meixner coefficients for the pulse response are determined and used to reconstitute the pulse response the result is extremely sensitive to errors in the data and the results for the realisation used as an example in this paper is worse than for any method except forward substitution. If, however, the Meixner series used as an expansion of the pulse response is limited to a small number of terms then excellent results can be obtained. The result for the case where twenty-five Meixner coefficients are obtained for the input and the output and then used to determine five Meixner coefficients of the pulse response by least squares is shown on Fig. 12.

It is clear from Table III that harmonic analysis (9 terms) and Meixner analysis (5 terms) give substantially better estimates of the pulse response than any of the methods previously examined. As in the case of the methods of constrained optimisation there is an element of subjectivity since no objective method of determining the degree of truncation has been developed for the field situation where no information is available concerning the nature or intensity of the error.

TABLE III

Results for Transform Methods

Method of identification	Mean r.m.s. error for 100 realisations	
	reconstituted output	estimated response
z-transform (polynomial division)	$9·2 \times 10^2$	$4·7 \times 10^3$
z-transform (root matching)	11·4	6·86
Harmonic analysis (32 terms)	0·091	0·562
Harmonic analysis (9 terms)	0·121	0·156
Meixner analysis (25 terms)	6·03	22·6
Meixner analysis (5 terms)	0·126	0·113

SYSTEM IDENTIFICATION USING CONCEPTUAL MODELS

This section deals with what were described earlier as conceptual models, which represent an approach to the analysis of hydrologic systems intermediate between black-box analysis described above and the solution of the differential equations developed in physical hydrology for the complex conditions of natural catchments. The explicit use of conceptual models in hydrology can probably be dated from the paper by Sugawara and Maruyama (1956). Starting from the observation that the recession curves of many rivers can be approximated by negative exponential functions, the authors suggested that such a function could be used to present the unit hydrograph and consequently that the system could be considered as equivalent to a hydraulic system in the form of an open vessel discharging through a capillary tube at the bottom, thus giving a linear relationship between the outflow from the vessel and the storage in it. They then attempted to improve the representation by modelling the behaviour of various rivers by a conceptual model in which several different vessels with different storage constants were arranged in parallel and the inflow divided between them. These tank-type models were further developed by placing the capillary outlet at a level higher than the bottom of the vessel, thus simulating a threshold effect of initial storage satisfaction, and also by tapping vessels by capillaries at different heights in order to produce a piece-wise linear storage discharge relationship that could approximate a non-linear relationship between storage and discharge.

Shortly afterwards Nash (1959) showed that the impulse response of a cascade of equal linear reservoirs was a gamma distribution and suggested that such a model could be fitted to unit hydrographs and the number of reservoirs in the cascade (n) and the storage delay time (K) of each reservoir determined by matching the first and second moments of the conceptual model to the actual first and second moments of the derived unit hydrograph.

In 1959 I suggested a more general conceptual model consisting not only of linear storage elements but also linear distortionless channels which would have the effect of applying a pure translation to an inflow (Dooge, 1959). All of the above conceptual models are clearly models of lumped systems, which are of the type considered here. It is of course also possible to develop conceptual models of distributed systems. The distributed model most commonly used in the analysis of hydrologic systems of both surface and subsurface flow is that corresponding to the convective-diffusion equation.

It is convenient to characterise the shape of the system response, whether derived in the field or based on a conceptual model, by means of the moments or cumulants of the system response with respect to time. In theoretical studies, advantage is taken of the fact that the Fourier transform of the impulse response is the generating function for the moments and the logarithm of the Fourier transform is the generating function for the cumulants. It can be shown that, for a linear time-invariant system, any specified cumulant of the output is equal to the sum of the corresponding cumulants of the input and the impulse response.

$$k_R(y) = k_R(x) + k_R(h). \tag{23}$$

Thus the estimates of the cumulants of the impulse response can be obtained from the computed cumulants of input and output for any given input-output event. Because of the relationship between moments and cumulants (see Kendall & Stuart (1958)), the relationship given by equation (23) holds for the first moment about the origin and for the second and third moment about the centre.

Numerical experimentation was carried out for the same data as before and a number of conceptual models fitted by optimising the values of their parameters through moment matching. For this purpose use was made of the computer program PICOMO (parameter identification of conceptual models) developed by O'Kane and Dooge (1977). Typical results are shown in Table IV.

TABLE IV

Identification using conceptual models

Conceptual model	Parameters	Mean r.m.s. error for 100 realisations	
		reconstituted output	estimated response
Single reservoir	K	0·451	0·841
Triangle (1:3)	T	0·271	0·363
2 equal reservoirs	K	0·173	0·219
Routed triangle	T,K	0·200	0·359
Routed rectangle	T,K	0·182	0·252
n equal reservoirs	n,K	0·194	0·274

The variances of the results in Table IV lie in the range 10^{-2} to 10^{-3}. It can be seen that all of the 2-parameter conceptual models estimate the pulse response as well or better than the methods of black-box analysis based on matrix inversion or optimisation. The best result of all is obtained for the 1-parameter model of 2 equal reservoirs which might be thought surprising in view of the fact that the 2-parameter model of n equal reservoirs seeks to optimise the value of n. It must be remembered, however, that the latter optimisation is based on moment matching and that the results of Table IV are compared on the basis of r.m.s. error. The ranking would be expected to be similar but not necessarily identical.

The results for the Nash cascade for the realisation used previously are shown on Fig. 13. It will be observed that this simple conceptual model works better than most methods of black-box analysis even when the latter include implicit constraints and post-filtering. The use of a conceptual model ensures that the volume is equal to one, that the pulse response is smooth and in the present case that the response is unimodal and of the right general shape. For the case of a bimodal response the performance of a unimodal conceptual model would of course be less satisfactory and would not be able to compete with the transform methods. However, for most hydrologic systems conceptual models represent a powerful method of determining the system response.

There is an interesting relationship between black-box analysis and conceptual models based on linear reservoirs and linear channels. This is more easily demonstrated for the case of systems with continuous input and output data rather than the discrete case described above. For such a case the continuous convolution equation represented by equation (3) given earlier is represented in the transform domain by

$$Y(s) = X(s) \cdot H(s) \qquad (24)$$

so that the system function, or Laplace transform of the impulse response, $H(s)$ is given by

$$H(s) = \frac{Y(S)}{X(s)} . \qquad (25)$$

If it is assumed that the system function can be adequately represented by a rational function, then we have

$$H(s) = \frac{P_m(s)}{Q_n(s)} \qquad (26)$$

where $P(s)$ and $Q(s)$ are polynomials. For the system to be stable, the degree (n) of the polynomial in the denominator $Q_n(s)$ must be equal to or greater than the degree (m) of the polynomial in the numerator $P_m(s)$. For a heavily damped system the n roots of the polynomial Q_n will be all negative and real.

Insertion of equation (26) in equation (24) and multiplication by $Q_n(s)$ gives

$$Q_n(s) \cdot Y(s) = P_m(s) \cdot X(s). \qquad (27)$$

Because $Q_n(s)$ and $P_m(s)$ are polynomials, inversion of equation (27) to the time domain gives a linear differential equation with constant coefficients of order n. This can be written as

$$\prod_{i=1}^{n} (1 + K_i D) \left[y(t) \right] = \prod_{j=1}^{m} (1 + L_i D) \left[x(t) \right] \qquad (28)$$

where K_i and L_i are equal to minus the reciprocals of the roots of the polynomials Q_n and P_m, respectively, and D is the differential operator. It is easy to show that equation (28) is the governing differential equation for a cascade of n linear reservoirs where the storage delay times are given by $K_1, K_2, \ldots, K_n$ and where the manner in which the input is divided up as lateral inflow into the respective reservoirs is controlled by the values of $L_1, L_2, \ldots, L_m$.

For the special case where the values of L_i are all zero, i.e., where the system function can be represented as the simple reciprocal of a polynomial Q_n, the system function corresponds to the case where all the inflow is at the upstream end. For such upstream inflow, it can be shown (by means of Jansen's inequality) that:

$$m_3 \leq (m_2)^{3/2} \qquad (29)$$

where m_2 is the dimensionless second moment and m_3 is the dimensionless third moment. This is equivalent to saying that the shape of the response function $K_1, K_2, \ldots, K_n$ is closely bounded on one side by the simple two-parameter model consisting of a single linear channel of delay time T combined with a single linear reservoir of storage delay time K. It can also be shown (by means of Holder's inequality) that:

$$m_2^2 \leq m_3 \qquad (30)$$

i.e., that the shape is closely bounded on the other side by the Nash cascade consisting of n equal linear reservoirs each of storage delay time K. Thus a model with an arbitrary number of parameters is closely bounded on each side by a 2-parameter model.

GENERAL COMPARISON OF METHODS

A general over-all comparison of the methods discussed above is shown in Table V. This shows not only the r.m.s. error of the reconstituted output and the r.m.s. error of the estimate pulse response but also the approximate c.p.u. time involved for each method on the IBM 360/50 system of University College Dublin.

As can be seen from Table V the improved estimation of the pulse response obtained by constrained optimisation involves a considerable increase in computer time. Recalling that the basic problem is to determine the underlying true pulse response, the numerical experimentation described above would indicate that for the particular case studied the most efficient methods are:

        Harmonic analysis

        Meixner analysis

        Conceptual models

Table V indicates that it is more economical in terms of computer time to accomplish the necessary smoothing of the pulse response by truncation of an orthogonal series

TABLE V

Over-all comparison of methods

Method of identification	Mean r.m.s. error of 100 realisations		c.p.u. time seconds
	reconstituted output	estimated response	
Forward substitution	$9 \cdot 2 \times 10^2$	$4 \cdot 7 \times 10^3$	0·1
Collins method	$9 \cdot 5 \times 10^{-2}$	0·557	0·5
Least squares	0·082	0·548	0·9
Regularisation	0·227	0·316	7·4
Quadratic programming	0·102	0·294	59·4
Root matching	11·4	6·9	3·7
Harmonic (F.F.T. 9)	0·121	0·156	0·3
Meixner (L.S. 25/5)	0·126	0·113	1·4
Conceptual models (n equal reservoirs)	0·194	0·274	0·8

expansion or by use of a smooth conceptual model than by introducing a smoothing constraint into the solution process.

In the use of transform methods, there remain the important problems (a) of choosing the optimum type of transformation for a given type of system and (b) of determining the optimum degree of truncation for a given set of data. In the case of conceptual models there is the key problem of choice of model and the further problem of parameter optimisation. Extensive numerical experimentation could suggest answers to these problems but guidelines based on mathematical analysis would be much more satisfactory.

REFERENCES

Body, D.N. (1959), "Flood Estimation: Unit Graph Procedures Utilising a High-Speed Digital Computer", *Water Research Foundation of Australia (Sidney) Bulletin*, **4**, p. 41.

Bruen, M. (1977) "A Comparison of Some Methods of Linear System Identification", unpublished report, Department of Civil Engineers, University College, Dublin, November 1977.

Chow, V.T. (Editor), (1964), "Handbook of Applied Hydrology", McGraw-Hill, New York.

Collins, W.T. (1959), "Runoff Distribution Graph from Precipitation Occurring in more than one time unit", *Civil Eng.*, **9**, p. 559-, New York.

de Laine, R.J. (1970), "Deriving the Unit Hydrograph without using Rainfall Data", *J. Hydrology*, **10**, pp. 379-390.

Dooge, J.C.I. (1959), "A General Theory of the Unit Hydrograph", *J. Geophysical Res.*, **64**, No. 2, pp. 241-256.

Dooge, J.C.I. (1965), "Analysis of Linear Systems by Means of Laguerre Functions", *SIAM J. Control*, A, **2**, No. 3, pp. 396-408.

Dooge, J.C.I. (1966), "Response of Heavily Discrete Damped Systems", Unpublished Memorandum, Department of Civil Engineering, University College Cork.

Dooge, J.C.I. (1973), "Linear Theory of Hydrologic Systems", *Agricultural Research Service, Technical Bulletin No. 1468*, U.S. Department of Agriculture, October 1973.

Eagleson, P.S. (1969), "Dynamic Hydrology", McGraw-Hill, New York, p. 462.

Freeze, R.A. (1972), "Role of Subsurface Flow in Generating Surface Runoff", 1. Baseflow contributions to channel flow, *Water Resources Res.*, **8**, No. 3, pp. 609-623.

Garvey, B.J. (1972), "The Analysis of Linear Systems by means of Laguerre and Meixner Functions", Thesis for M.Eng.Sc. degree, University College Cork.

Kendall, M.G. and Stuart, A. (1958), "The Advanced Theory of Statistics", Vol. 1, Chapter 3, Moments and cumulants.

Ku L.S. (1967), "Solution of Inverse Problems for Linear Flow Problems", *Soviet Hydrology*, 2, pp. 194-199.

Laurenson, E.M. and O'Donnell, T. (1969), "Data Error Effects on Unit Hydrograph Derivation", *J. Hydraulics Division ASCE*, 95, No. HY6, pp. 1899-1917.

Nash, J.E. (1957), "The Form of the Instantaneous Unit Hydrograph", Proceedings of Toronto General Assembly of IASH, Vol. 3, Surface Water and Evaporation, pp. 114-121, IASH Publication No. 45.

Newton, D.W. and Vinyard, J.W. (1967), "Computer-Determined Unit Hydrographs from Floods", *J. Hydraulics Division ASCE*, 93, No. HY5, pp. 219-235.

O'Donnell, T. (1960), "Instantaneous Unit Hydrograph Derivation by Harmonic Analysis", IASH General Assembly of Helsinki, IAHS Publication No. 51, pp. 546-557.

O'Kane, J.P.J. and Dooge, J.C.I. (1977), "PICOMO: A Program for the Identification of Conceptual Models", Workshop on Mathematical Models in Hydrology, Pisa, December 1974. Published in Ciriani, Marione and Wallis (editors). "Mathematical models for surface water hydrology", Wiley-Interscience, London.

Snyder, W.M. (1955), "Hydrograph Analysis by the Method of Least Squares", *J. Hydraulics Division ASCE*, 81, pp. 01-25.

Sugawara, M. and Maruyama, F. (1957), "A Method of Prevision of the River Discharge by Means of a Rainfall Model Darcy Symposia", Vol. 3, Floods, IASH Publication No. 42, pp. 71-76.

Natale, L. and Todini, E. (1973) "Black-box Identification of a Flood Wave Propagation Linear Model", Proceedings of Istanbul Congress Vol. 5, pp. 165-168, International Association for Hydraulic Research.

Wolfe, P. (1959) "The Simplex Method for Quadratic Programming", *Econometrica*, pp. 382-398.

OPERATIONAL ASPECTS OF WATER RESOURCE PROBLEMS

J.C. Wilkinson

(University of Lancaster)

It seems desirable to begin by indicating what the word "operational" means to me. I interpret it in the same way as it is used in my chosen discipline, operational research. This to me means research into *operations*, of whatever kind, but it does, I believe, imply a more direct relevance to problems in the real world than is the case with what may be loosely termed pure research or abstract research. This is not to deny the value of the latter nor to suggest that significant progress on the solution of real-world problems is possible without a good number of assumptions and approximations being made; indeed the characteristic approach of operational research is the construction of a model, which of itself necessitates simplification. It is to suggest that the "abstract" model on which the analyst works is interpreted literally to mean a model *abstracted* from a real situation. The material in this paper is, therefore, taken from practical studies in which I personally have been, or am, involved. I am not attempting a review, or "state of the art", paper.

1. Let me start by reflecting upon some points that emerged, problems that were encountered, and solutions that were devised, during the course of the River Dee Regulation Research Programme, which provided my first involvement with problems of water resources, some 10 years ago.

The Dee is a river in North Wales flowing eastwards some 70 miles from Llyn Tegid (more commonly called Bala Lake by Englishmen) through the city of Chester and then into the Irish Sea (see Fig. 1). Since 1956 the outflow of Llyn Tegid has been controlled by sluice gates, and since 1965 a further, larger reservoir, Llyn Celyn, has been added to the system to provide additional water supplies for the benefit of the city of Liverpool. Except in periods of very high precipitation the Celyn/Tegid complex approximates a single, multi-purpose, regulating reservoir (see Fig. 2). The scheme provides

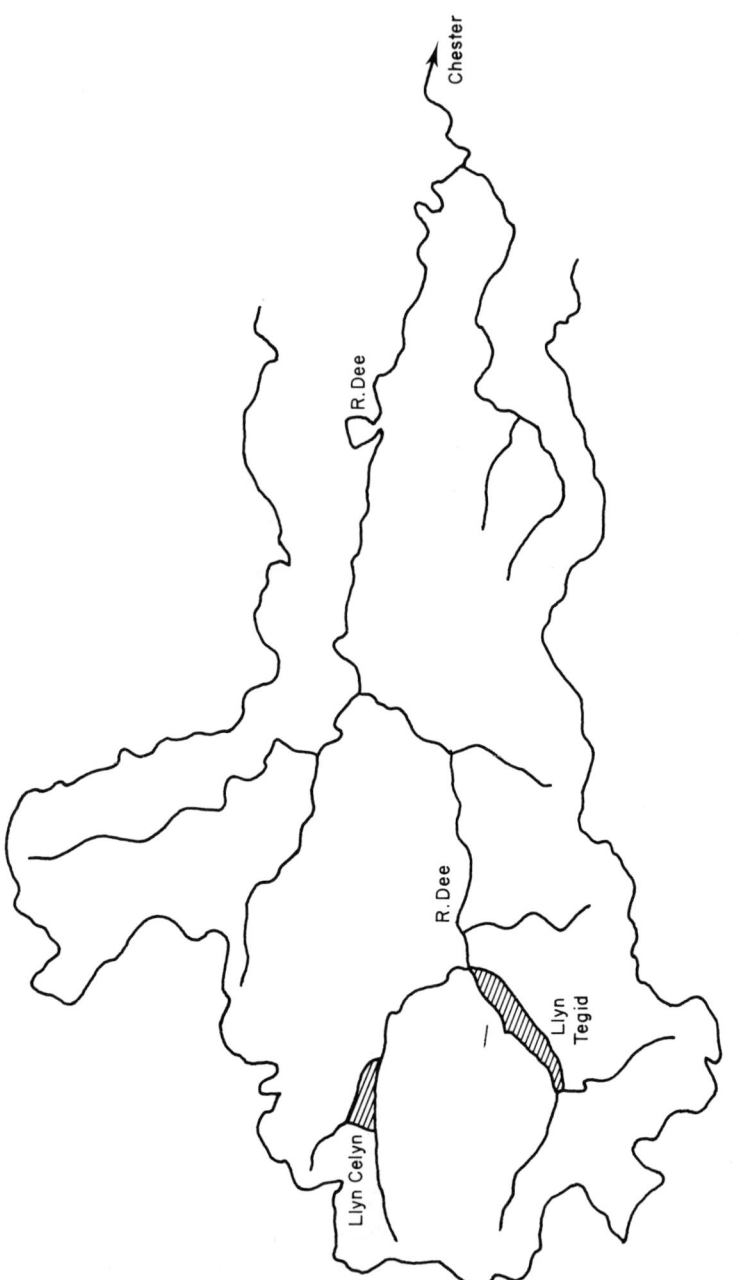

Fig. 1

benefits of water supply, flood mitigation, good fishing, recreation (both on Bala Lake and in the downstream Dee) and some hydro-electric power among others.

Of these benefits it was claimed that the first two, namely water supply and flood mitigation, were of infinitely more importance than any others and I was asked to investigate the optimal operating of the scheme with reference to these two considerations only, giving water supply absolute priority over flood mitigation. I do not intend to present the full story of how the problem was approached. It is sufficient for my present purpose to note that, largely because of the different time scales over which droughts and floods occur, two distinct phases were recognised - long term control and short-term control.

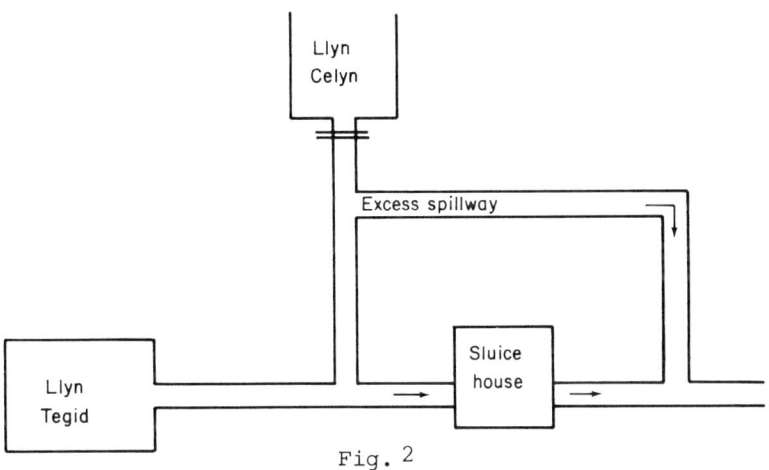

Fig. 2

Long-Term Control - if one is concerned only with the single objective of maximising water supply the solution is simple; keep reservoirs as full as possible and release only as much as is necessary to maintain desired yield. To incorporate the objective of flood mitigation I sought to find reservoir levels as low as possible (in order to maximise potential flood mitigation) subject to satisfying a pre-specified reliability of water supply. This immediately brought me into the area of relability definitions which I shall consider later. In the event I sought retention levels, seasonally varying of course, such that the return period of a failure was 25 (and later 50) years. But failure was undefined! It was agreed the this was to mean that no more water could be supplied by the

reservoirs, but there was no agreement on how much water in
the reservoirs was usable, nor was any rationing policy established for controlling the rate of reservoir drawn-down in
times of imminent shortage. At the same time it was confidently claimed that the current operating rules guaranteed
protection against the "one in fifty years" drought.

Results of long-term study and comparison with existing
practice - the initial phases of this work are described in
some detail in Wilkinson (1972). Two major lessons are to
be learnt from it.

(1) Intuition is not reliable in finding solutions to problems
of this nature; the complicated interactions of a sequence of
probability distributions (of successive monthly runoff
figures) are too difficult to comprehend without analysis.
The current practice of maintaining high retention levels in
summer and low in winter seem at first to accord with the
desired objective. Analytical study, however, suggested the
opposite (see Fig. 3) because the probability distributions
of runoff in summer and winter are such that if reservoir
storage is not higher than current practice allows in February
and March there is a real chance of failure in summer, and
winter runoff so exceeds demand that high retention levels
in the summer months are unnecessary for supply purposes.

(2) Management's objectives are more apparent from their
actions than from their words. The recommended retention
levels, with respect to the stated objectives, are unacceptable in practice because in practice the true objectives
differ from those stated. Current operating levels are high
in summer because management believes that the chance of
flood-producing precipitation in summer is negligible and
chooses to give higher priority to another, unstated, objective
namely amenity value. No-one, least of all holiday-makers
and those whose living is dependent upon holiday-makers, likes
an empty lake. In winter it is clear that operating levels
are lower in order to provide extra flood protection. This
policy is equivalent to attending to that disaster which is
most imminent, which may be justifiable but it is important
to realise that increasing protection against winter floods
decreases protection against drought the following summer
and such decisions ought not to be taken without some assessment of the relative costs of different disasters.

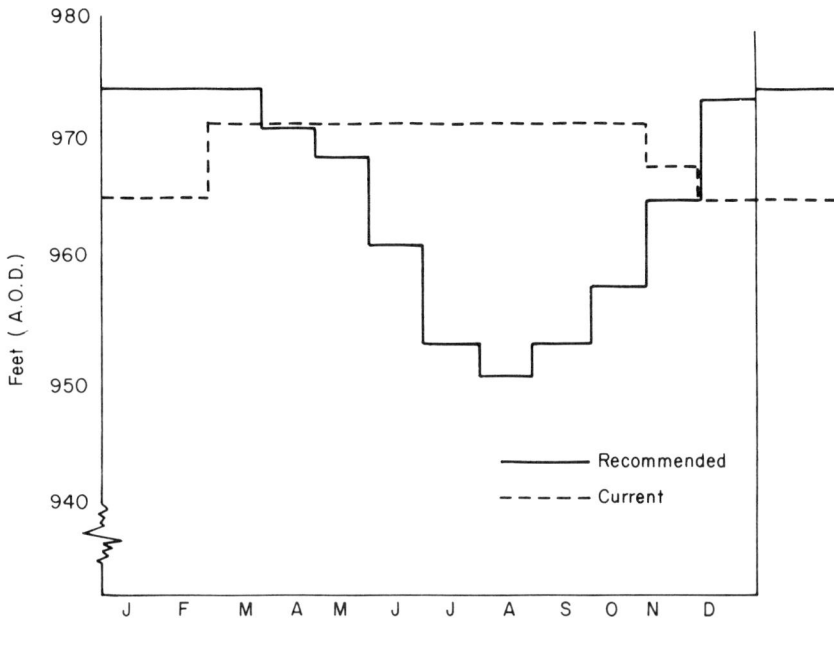

Fig. 3

Short-Term Control - Here I was concerned entirely with the threat of floods but the problem has again multiple objectives, namely to avoid floods in different places. The context of the problem is more fully described in Jamieson and Wilkinson (1972).

Essentially one has, at any time, knowledge of the lake level in Llyn Tegid, the river level in the Dee at a gauging station, and the current rate of release of water into the river from the lake through controllable sluice gates. In addition, one has available, through the integration of sub-catchment models and a complex telemetry system covering the whole catchment, forecast hydrographs of input into Tegid and into the river between Tegid and the downstream gauging station for some n time-periods ahead (see Fig. 4). If one denotes the lake and river levels by a and b, the current release by y, and hydrographs by $H_A(n)$ and $H_B(n)$ and associates "costs" of flooding $A(a)$, $B(b)$, respectively, with the lake and river levels, then one can formulate the problem of minimising the maximum flooding cost incurred during the next n time intervals in dynamic programming terms as follows.

$$f_n(a,b,y) = \min_{x} \left\{ \max \left[A(a), B(b), f_{n-1}(T(a,b,y|x)) \right] \right\}$$

$$= \max \left[A(a), B(b), \min_{x} f_{n-1}(T(a,b,y|x)) \right]$$

with $f_0(a,b,y) = \max \left[A(a), B(b) \right]$, all y.

T: $y = y + x$
$a = a - y + H_A(n + \text{lag})$
$b = y + H_B(n)$

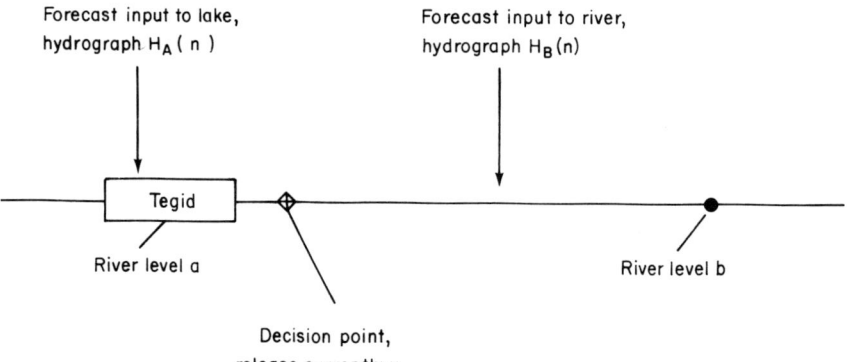

Fig. 4

In this formulation the time of travel of water released from Tegid to the gauging station is assumed constant and denoted by lag in the transformation equations. The decision variable x is the increase in the rate of release.

In reality the forecast hydrographs are subject to error so that a stochastic dynamic programming formulation is more properly required. But even this would not remove the uncertainties in establishing appropriate cost functions A(a) and B(b). Strictly these should take the value zero as long as a,b are below the maximum capacities of the lake or river. If, however, one specifies A and B of the form in Fig. 5, then one can, I believe, take some account of the fact that the hydrographs contain forecasting errors, enable the problem to be solved (albeit approximately) by deterministic rather than stochastic dynamic programming, and simultaneously incorporate a trade-off between the costs of flooding at different places.

The computational time and storage requirements may be reduced by choosing time intervals and units of volume such that the only permissible values for the decision variable x are -1, 0 and +1, and with little or no loss of realism these may be reduced to ± 1 provided the time increment is kept small. This results in a significant reduction in computational requirements since the evaluation of each $f_n(a,b,y)$ then needs only two calls of the procedure T and a single comparison, and the optimal decision x can be stored simultaneously with the optimal value by storing the product $x \cdot f_n(a,b,y)$.

Having simplified the problem to the point where the decision variable at each instant is ±1 it is now possible to suggest a computational procedure which avoids the need for dynamic programming and enables evaluation of the optimal decision to be calculated on-line. Fig. 6 is a representation of the situation faced at each decision instant. Any sequence of decisions may be represented as a continuous path from the initial point along arcs of the network. It is also clear that the path moving ever upwards, corresponding to the decisions (+1, +1, +1, ...), represents that policy which minimises the flooding cost at the lake and maximises the flooding cost at the downstream gauging station. [The sequence (-1, -1, -1, ...) does precisely the opposite.] For any policy it is a simple and very quick procedure to establish whether the maximum flooding cost over the duration of the forecast hydrographs is incurred at the lake or downstream, together with the node on the representative path at which this occurs. If, on the path (+1, +1, +1, ...), the maximum cost is incurred at the lake then nothing further can be done to reduce this cost. If, on the other hand, the maximum cost is incurred

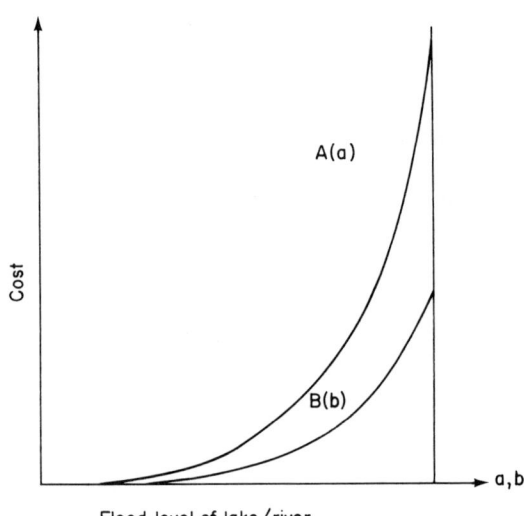

Fig. 5

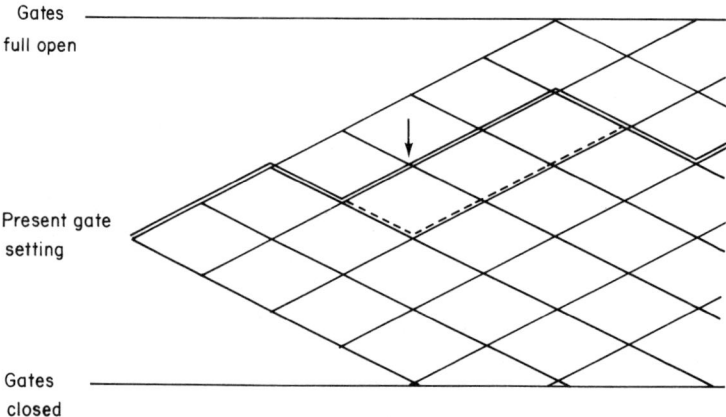

Fig. 6

downstream, at node i say, then the path may be perturbed by replacing the +1 decision at node (i-1) by -1. This modified policy may then be evaluated and the process repeated until the minimum cost is incurred, if ever, at the lake. In general the node at which the maximum flooding cost (downstream) occurs is found and the path is then replaced by that path which replaces +1 by -1 at the latest possible node before the node in question; the subsequent decisions in the revised policy are +1 subject to its being bounded above by the existing policy. Fig. 6 shows a typical modification of a policy.

It may be shown that if a policy spans n time intervals the maximum number of iterations to obtain the optimal policy is $\frac{1}{2} n(n + 1)$; it will usually be very much smaller. With computational effort reduced to this there is no difficulty in incorporating these calculations into an on-line procedure to be repeated each time a real-time decision is required, perhaps every half-hour.

2. I should now like to turn to a problem which is under current investigation. Consequently in connection with this I shall be raising more questions than I shall be providing answers. The problem exists in a small catchment in the South of England where demand is met normally by abstraction from a river but by augmentation of this from a sandstone aquifer if the natural flow of the river is inadequate. The rate of natural recharge of the aquifer is less than the average rate of abstraction to meet demand and consequently there has been constructed the facility for artificially recharging the aquifer at times of surplus flow in the river (see Fig. 7).

It is commonly assumed that the demand for water in Britain will greatly increase over the next 25 years and this catchment provides no exception. Consequently the facilities for abstraction from the aquifer and for recharge, together with the associated treatment works, must be increased, at considerable capital expenditure, to meet the steadily increasing demand. The problem, therefore is to devise an optimal investment schedule over the next 25 years incorporating an optimal operating strategy.

It should be noted that this problem embraces both design and operating aspects, and that they interact. It is certainly possible for example that increased investment in treatment plant capacity might lead to reduced day-to-day operating expenditure by avoiding the need for pumping at peak tariffs. Nor is the decision on the optimal combination trivial.

Consider two alternative systems, each equally feasible. The first system involves a capital cost A and an annual cost B. The second involves capital cost A' and annual operating cost B'. What should one do if A < A' and B > B'?

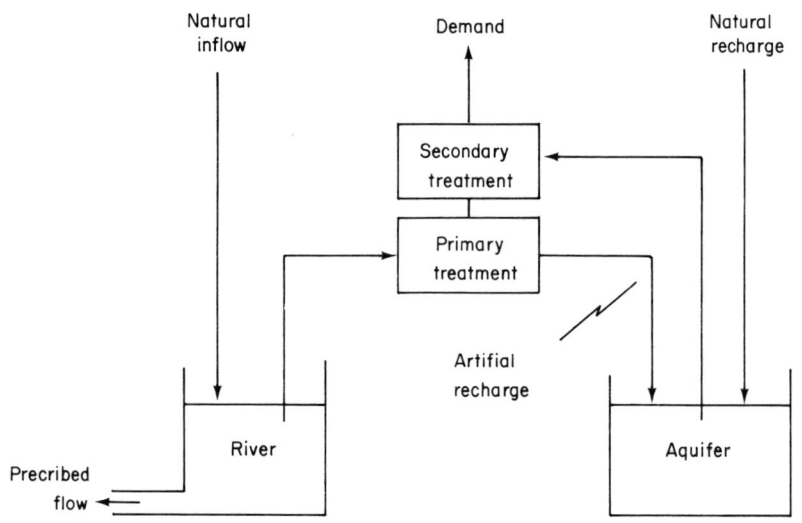

Fig. 7

Discounted cash flow theory suggests that one should compare total present cost. For a time horizon of n years and discount factor (assumed constant) of α one has the following costs of the two systems.

First: $\quad A + B (1 + \alpha + \alpha^2 + \ldots + \alpha^{n-1})$

Second: $\quad A' + B' (1 + \alpha + \alpha^2 + \ldots + \alpha^{n-1})$

and the choice depends very much upon the (largely arbitrary) choice of values for α and n. Nor does it help very much to suggest that one should attempt to minimise cost over an infinite horizon. Although in the discounted cash flow model

Operational Aspects

this eliminates one of two parameters, in the context of the real problem one must beware of making the assumption, albeit implicitly, of no further growth in demand beyond the next 25 years. And, in any case, after the experience of recent years the one thing about which one can be reasonably certain is that α will not be constant.

In considering the combined investment/operating problem, making such acts of faith in our forecasts of the future as are necessary to proceed, an integrated model, incorporating both investment and operating decisions, would seem to be indicated; and the nature of the problem - multistage with today's decisions having a very definite sequential influence upon tomorrow's problem - would suggest dynamic programming as a likely tool for the job. However, the prospect of solving such a problem with multidimensional state and decision variables and stochastic transformation operator is not attractive. If one also considers that investment decisions are made on a year-to-year basis whereas the decisions about recharge are made from day to day, the separation of the two aspects of the problem seems very advisable.

Consider a river flow sequence. Demand and recharge capacity establish bounds such that if flow is below the one augmentation of supply from the aquifer is required, and if flow is between the two bounds then recharge is possible (see Fig. 8). For known demand and recharge capacity it is possible to estimate the joint probability distribution of this demand on, and potential recharge of, the aquifer on, say, an annual basis. Of course the demand and potential recharge will be correlated with each other and also with the annual rainfall and hence the natural recharge.

If natural recharge is x, potential artificial recharge is y, and aquifer draw-down to meet water supply is z, then one might, at the beginning of year n from a finite horizon, formulate the minimum expected cost function,

$$f_n(s) = \min_r \left[C_1(s,z_n) + C_2(r) + \int\int\int_{x,y,z} f_{n-1}(s-z_n+r+x) \phi_n(x,y,z) \, dxdydz \right]$$

where s is aquifer storage at the beginning of year n, r is the amount of artificial recharge subject to constraints imposed by the potential artificial recharge, y, and the current

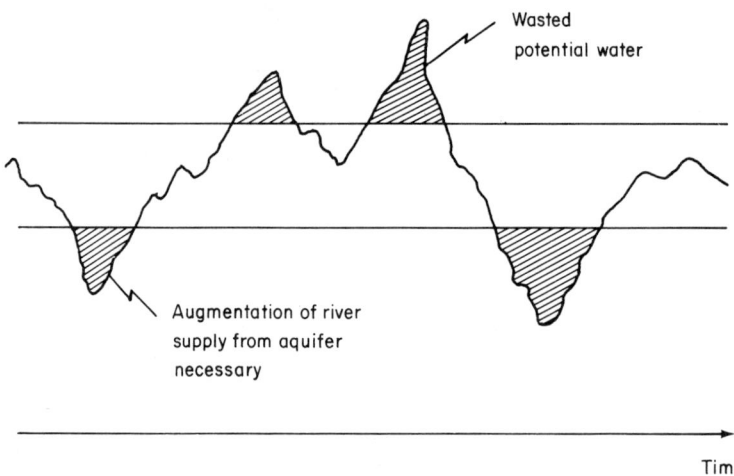

Fig. 8

recharge capacity. C_1, C_2 are, respectively, the costs of drawing down and recharging the aquifer and ϕ is the multivariate probability density function (p.d.f.) of x, y, z. The minimisation is also subject to constraints, so far unspecified, on long term reliability of supply. Note that the p.d.f. ϕ varies with n since the p.d.f. depends upon the demand and recharge capacity in year n.

The solution of such a dynamic programming problem is still a formidable task. One should ask whether it is necessary. If demand and recharge capacity were fixed and known to be the same each year then the operating recharge problem would reduce to that of finding the best target aquifer level to try to attain. It may vary throughout the year but this is not necessarily the case. The cost of having this level too low would be reflected in shortages of supply occurring too frequently. The cost of setting the target level too high would result in excessive expenditure - incurred by artificially recharging the aquifer at a time when it would be recharged naturally anyway. Such a situation is analogous to the classical "newsboy" problem of operational research and has been extensively studied.

Operational Aspects 51

If the operating problem is, albeit sub-optimally, solved in this manner, the more important investment problem assumes a simpler format. But questions still remain to be answered. If, as seems sensible to me, a dynamic programming formulation is adopted, there is a dilemma between minimising total (present) cost over an infinite future, with the difficulties of forecasting inherent in that, and minimising cost up to, but not beyond, the planning horizon. If the latter alternative is chosen then it is not unlikely that the recharge and treatment capacities in the system in the year 2001 will be woefully inadequate to meet the needs of the year 2002 - and the aquifer will be almost empty.

Another possibility is open to us.

(1) On the assumption that demand is constant, at the level of the year 2001, for every year thereafter, find - possibly using the policy iteration method of dynamic programming - the best levels of recharge and treatment capacity to meet this demand.

(2) Accepting these levels as terminal states 25 years hence, and knowing the corresponding present states and growth of demand in the intervening period, find the optimal (= least cost, subject to reliability of supply) path from now to this horizon.

I would expect a simple perturbation technique, similar to that used in the short-term control problem on the Dee, to be more efficient than dynamic programming for solving this problem.

3. I shall now devote some attention to the question of <u>reliability</u>. It seems to me that this is a term which, perhaps more frequently than any other, is abused in discussing water resources problems. Rarely is a precise definition attached to the term. Few practising engineers would claim that a system designed to fail only once in 10 years is certain to fail in 1977 if the last failure was in 1967 - failures are statistical or stochastic phenomena, they will assert - but how does common practice accord with this assertion? Systems are still designed and constrcuted to withstand a "design" drought or "design" storm, whatever that is meant to imply. It is not unusual to construct a synthetic drought sequence by concatenating the driest historical summer, the driest historical winter, and the second driest historical summer, and demanding that the system being designed should not fail

under such experience. No attempt is made to estimate the expected frequency of occurrence of this rare sequence of events. Again, the 100-year extreme event is frequently estimated by plotting cumulatively the extreme annual events in the historical data, usually on logarithmic paper to make them fall more nearly on a straight line, and then projecting this line deterministically (perhaps using least squares) to give the once-in-a-100 years estimate.

Things are sometimes even worse. I was suprised to learn, in connection with a study on the design of storm sewers recently, that design storms are traditionally assumed to have a particular, and to me unrealistic, distribution over time.

One might ask whether water resource engineers really benefit from the increasing use of synthetic hydrology for design purposes. One is frequently interested in designing a system to provide protection against failure measured on an annual basis. Perhaps 5 to 10 years' daily data exist. Certainly the annual statistics from 5 to 10 years are very few, so what is often done? A daily model is postulated and fitted to the daily data. Synthetic data, perhaps 500 years or more, are then generated and the annual statistics from these synthetic data are subsequently used for design purposes, as if these are representative of the assumed underlying population. The idea persists that one can, by generating synthetic sequences of data, get information for nothing. But if, as is often the case, one is interested in extreme events, or the year-to-year correlation structure of the phenomenon under study, the fact remains that the original data are too sparse to test properly that these statistical characteristics are representative of the real world, rather than of the model itself.

Perhaps it would be more honest to construct an annual model possessing those characteristics which one believes, from one's own or others' experience in different contexts, the population might reasonably possess, and use the historical records merely to check consistency in respect of those statistics for which they provide a large enough sample.

In speaking of rare events such as, hopefully, floods and droughts, one commonly uses the term "return period" to describe the average time between successive occurrences. In the absence of definite information to the contrary it is usually assumed that such failures occur at random, the times between successive failures being distributed negative-exponentially. Indeed the time to the next failure from any randomly chosen instant is negative-exponentially distributed, the probability of the next failure occurring later than t from that instant being

$$p(t) = e^{-t/T}$$

where T is the return period according to the definition just given.

This equivalence between p and T was used in formulating a recursive model in the Dee Research Programme relating the probability of no shortage in the next n months to the probability of no shortage in n-1 months.

The assumption of negative-exponential inter-failure times is equivalent to assuming that the number of failures occurring in a fixed time interval is a Poisson variable with mean proportional to the length of the interval. It is also equivalent to assuming that the probability of a failure in any small interval is time-homogeneous. In a practical situation this is unlikely to be realistic, not only because of seasonal effects, but because, for example, a reservoir requires a non-negligible time to recover from a shortage and consequently the probability of another shortage during this recovery period is greater than during a randomly chosen period.

The danger of using these measures as criteria of reliability has been illustrated by, among others, Jamieson, Radford and Sexton (1974). Instead they advocate use of the measure "cumulative percentage frequency" of some specified failure level being transgressed.

This certainly is an improvement on existing common practice but does it go far enough? It still leaves open the question of definition of failure. At one extreme this might be taken as a catastrophic failure - corresponding to there being absolutely no water available, while at the other it may indicate merely that a reservoir is drawn down a metre or so below its target retention level for the time of year.

If one is interested in making probabilistic statements about a reservoir reaching any level below that at which corrective action is first taken (I am thinking particularly of imposing some form of restriction on permitted abstractions) then one must know what effect these "rationing" decisions will have. It is the unfortunate truth that, for most supply systems in the UK, the storage levels at which different degress of rationing these should be, are not specified.

The situation has been highlighted by the emergency bill rushed through Parliament this summer (1976) at the instigation of the regional water authorities to give them increased and more flexible power in imposing restriction on the use of water in order to deal with the severe water shortage experienced in many parts of the country.

I would advocate an extension of the concept of "cumulative percentage frequency" beyond application to a single specified failure level to calculating this statistic with reference to the transgression of several different levels of interest, appropriately those levels at which successively severe degrees of rationing are imposed, thereby reducing the frequency with which the next level would be reached without the imposition of rationing. There is also an attraction about the idea of describing reliability of supply in this way: for any fraction of requested supply one would know the probability (= reliability) of being able to meet or exceed that fraction.

Let $f(x)$ be the probability (density) that a reservoir contains a proportion x of its capacity, let
$r(x)$ be the permitted abstraction when reservoir level is x, and let
$p(z)$ be the probability that input to the reservoir in a unit time interval is a proportion z of its capacity. Then, assuming successively independent inputs (this clearly depends upon the choice of time interval)

$$f(y) = \int_0^1 f(x) \cdot p(y - x + r(x)) \, dx.$$

For a given rationing curve $r(x)$ we can readily evaluate the corresponding reliability curve $f(x)$ (see Fig. 9 for typical curves), and with more work we should be able to specify a minimally acceptable reliability curve $f(x)$ and find a corresponding rationing curve $r(x)$. I say a corresponding rationing curve because it is not necessarily unique.

In this paper I have attempted to outline some of the problems and possible approaches in three studies, one of some years ago, one of current interest and one which I believe merits closer attention in the immediate future. I have drawn attention to the difficulty in producing results which were readily implemented. On reflection it seems to me that much of this difficulty arises from researchers either addressing problems which they themselves find interesting rather than the decision-makers', or believing that the necessarily simplified models they were analysing truly represented the real decision process.

I would like to end by expressing the hope that more mathematicians and statisticians will take the trouble to acquaint themselves with problems in water resources as perceived by those responsible for taking the decisions and that those enlightened practising water engineers who wish to avail themselves of the analytical tools now available will give more careful and serious thought to their objectives and more precise definition of the terms they use.

Rationing
$r(x)$ = abstraction authorised as percentage of demand

Reliability
$1 - F(x)$ = probability that reservoir level exceeds x

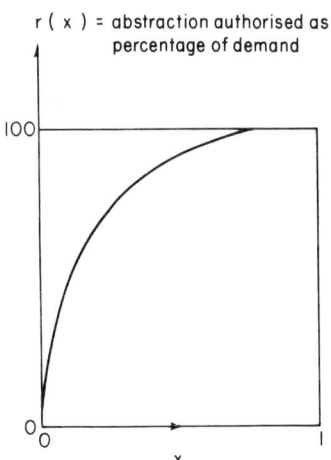
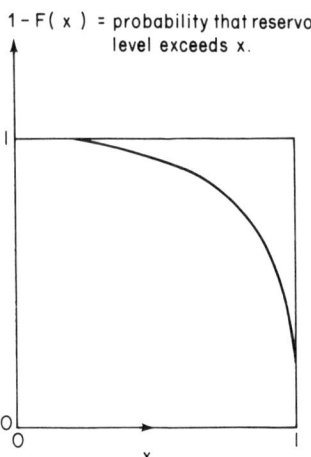

x is fraction of reservoir that is full

Fig. 9

REFERENCES

Jamieson, D.G., Radford, P.J. and Sexton, J.R. (1974), The Hydrological Design of Water-Resource Systems, Water Resources Board, Reading, U.K.

Jamieson, D.G. and Wilkinson, J.C.(1972) River Dee Research Program 3. A Short-Term Control Strategy for Multipurpose Reservoir Systems. *Ibid. Water Resources Research*, **8**, 4.

Wilkinson, J.C. (1972) River Dee Research Program 2. A Long-Term Control Strategy for a Multipurpose Reservoir. *Water Resources Research*, **8**, 4.

STOCHASTIC STORAGE PROBLEMS: THE WATER MANAGEMENT BACKGROUND

B. Rydz

(Severn-Trent Water Authority)

SYNOPSIS

In deciding what to do about water resource development and operation the engineer will try to use any tools which the applied mathematician puts at his disposal. But in the nature of things he will always be using his judgement to reach beyond strictly rational solutions. In dealing with the reliability of alternative patterns of water resource development the gap between rational solutions and the complex issues which have to be weighed in practice is uncomfortably wide. And as we create institutions with wider terms of reference, which are expected to recognise wider ranges of options, we naturally push the problems of optimisation further out of reach of rational solutions.

It has been accepted for many years that the standard of service we get from a water resource (in the sense of how well it copes with the vagaries of the climate) can best be expressed in probabilistic terms, and this was a considerable step forward. But the terms we use to specify and compare the vulnerability of quite different systems to failures of differing kinds are still very naive.

If some of the statistical sins of the past can be avoided by assuming a *priori* distributions only for certain natural phenomena and by converting these into the distributions of the consequences which really concern us, in the way indicated by Professor Lloyd, so much the better. But it seems that this approach has a long way to go to catch up with some of our typical problems.

It may be that by surveying the problems which are encountered in practice and the decision processes which are applied the mathematician will get a better idea of where the shoe really pinches and will be able to suggest more direct ways for the engineer to achieve his purposes, of which the most important, in this field, are:

(i) to achieve a consistency of approach to alternative ways of extending a resource system, and
(ii) to choose a performance standard which makes economic and social sense.

INTRODUCTION

This paper will contain no mathematics! What I would like to do, for mathematicians and others unfamiliar with this background, is to set the question of the adequacy of storage in the context of the decisions facing the "water manager" - by which I mean the group of people, lay and professional, responsible for running a modern water authority. I have in mind particularly the new water authorities in England and Wales which have a very wide range of functions, including the provision of all water supply and sewage disposal services, management of the water resources on which they are based and allocation of water rights to other users. Some of them operate systems in which groups of sources (reservoirs, river intakes, boreholes) are interlinked so as to serve a number of distribution networks in parallel and/or in series along a river system. Their options are therefore wider and their problems more complex than those of some water managements elsewhere. But I believe this form of management and these problems will be more widespread in the future.

Like most managers, the water manager runs a system bounded on one side by a set of resources and on the other by a set of customers. His job is to achieve a satisfactory balance between them, bearing in mind financial and environmental constraints. The authorities in this country can aim at an optimum deployment of resources over all services; that is what they were set up to do. This involves a lot of decisions about priorities and about the standards of service people want. The choice includes that of reliability: the ability of the system to continue providing an unrestricted service, despite shortage of rain or other vicissitudes.

Because flows in nature are variable and individually unpredictable, many performance standards in water management can be expressed only in stochastic or (more familiar word) probabilistic terms; in terms of what is to happen on the whole but is only more or less probable in any specific instance. The case in point here is the performance of a reservoired source, which is a combination of a variable natural flow and a store. The performance is usually described by a yield, or minimum flow available from the source, together with a measure of the reliability with which that yield is obtainable or its converse, the likelihood that the source will fail to deliver. More yield can be had with less reliability (or *vice versa*): more of both by tapping additional flows or adding more storage. Of the four variables, we usually try to fix the reliability according to a convention, optimise the storage for the available flow and determine the resulting yield. But in the case, for example, of the Derwent Valley reservoirs which are referred to later, storage volume, catchment inflow and the required reliability have been separately changed from time to time, with consequent adjustments to nominal yield.

DECISIONS ABOUT RELIABILITY

The need for decision usually arises in connection with the proposed introduction of a new source. The decisions are of two kinds and in this country both are usually tested at a public local inquiry or in private bill proceedings in Parliament:

(i) is it time something was done to augment resources in order to maintain the proper level of reliability? (proof of need);
(ii) if so, what should be done? (examination of alternatives).

The first requires an evaluation of the existing system of resources, based on the choice of a reliability criterion and a value on this reliability scale. The second depends upon examination of the merits of alternative new projects, including a consistent way of assessing their reliability or, to be more exact, the reliability of the system after they have been incorporated. Only the first is a question of absolute reliability (value); the second is one of relative merit. The answer to either question will depend in some degree on a third question:

(iii) how will the system be operated in practice? In particular, what will be the priority in exploiting individual sources, to what extent will their storage be shared out for maximum combined yield (known as conjunctive use) and what steps will be taken in the face of water shortage?

CHOICE OF RELIABILITY

A rational choice of reliability must depend on economic comparison of what it costs the water authority to buy more security and what it costs consumers to make do with less, but cost functions have rarely been available in this form. The metered consumer, if sufficiently informed, can make a direct cost comparison; but since most consumers of public water supply in this country do not buy the service by measure, measuring their preferences is a difficult problem in market research. One difficulty is that unless opportunities are taken to pursue enquiries at times of shortage, which in this country are few and far between, one is unlikely to get a balanced view. Until more economic analysis and research is done, the criterion will remain arbitrary; that is, it will be a matter of engineering conventions supported by considerable experience of how people react to the service they get and to the bills they are asked to pay. Their reactions may well change with time; for instance, people who until recently thought that our water services were too conservatively designed, or were using too much land, may now think differently because they have experienced deprivation. But there can be little point in exchanging one arbitrary criterion, to which managers and consumers are accustomed, for another.

The convention probably originated in the early years of modern reservoir engineering, about a century ago, when it was usual to give the yield of a direct supply impounding reservoir in terms of its catchment area and the average rainfall, with an allowance for evaporation loss, and empirical rules were developed for the sizing of the reservoir to give this normal or "safe" yield. In upland areas in Britain the yield was set at about 80% of mean rainfall (this being believed to be the average over the most severe sequence of 3 dry years), less an allowance for evaporation loss; and Parliament usually divided this yield arbitrarily between flow down the supply aqueduct and releases of compensation water to the catchment downstream. Some interesting observations on this practice are to be found in the evidence given by eminent engineers of the day to the Royal Commission on Water Supply just over a centry ago (1869), from which it is apparent that the notion of a definable risk of failure was only just beginning to enter their thinking.

Since then designers have considered a wider range of yields and storage volumes at any particular site, partly because they have been dealing with larger catchments and more capacious sites, but mainly because two other types of reservoir have come into common use. They are the pumped storage reservior, which is filled by pumping water from some conveniently placed river; and the regulating reservoir, which is emptied by making releases to a river to support abstraction at some remote point. Each associates the reservoir with a catchment larger than its own gravity catchment and enables a large volume of storage to be used more productively. Pumped storage regulating reservoirs also exist and are likely to become more common in future.

At the same time new ways of assessing the relationships between yield, storage and the security of the supply have been developed. In the first place historic records of flow were used to replace assumptions about rainfall and evaporation; where records are short or of doubtful quality attempts have been made to establish correlation with better flow records elsewhere or, by various forms of catchment modelling, with records of rainfall. Secondly, statistical theory has been invoked, making a virtue of the random nature of hydrological events - which had hitherto been at the root of the water engineer's problem - and so enabling him to quantify the reliability. The resulting problems (many arising, paradoxically enough, from the non-random aspects of these events, and more particularly of their consequences) are the main business of this chapter.

An attempt to survey the evolution of present methods of yield evaluation would be out of place here. A useful short survey and commentary on current methods has recently been compiled by the Central Water Planning Unit (1976).

The interesting point is that, in this country at least, engineers have sought throughout to rationalise, rather than to change, the reliability which the 19th century engineers, in their undoubted wisdom, thought was good for us. In recent years this has usually been labelled as a once in 50 years or once in 100 years liability to "failure." This is usually demonstrated in design by the retrospective simulation of a sequence of flows and reservoir operations in which a steady demand persists, unaffected by the imminence of failure, until the reservoir is just emptied, at which point relief arrives in the form of natural inflow. The demand which produces this result with inflows which are thought to have a mean recurrence interval of 50 (or 100) years is equated with the "safe yield". This quite unreal scenario has served as a model for the more complicated things which would happen in practice, when there would be no foresight of the ending of the drought and prudent operators would be obliged to introduce restrictions on water use or to curtail supplies. The shortcomings of the present approach and the possibility of more realistic criteria of failure will be referred to later, but meanwhile it is worth noting that a "design year" will normally be one in which restrictions are imposed because in practice there is no foresight of events; and that from the consumer's point of view, an authority's restriction policy will largely determine the impact of any "failure".

EVALUATION OF ALTERNATIVE PROJECTS

So much for the level of reliability (question *(i)* above), which should determine the timing of new projects. However arbitrary that may be we are still faced with the choice of project (question *(ii)*). This requires consistency of treatment so that the quantifiable elements can be reduced to some cost criterion - say, a discounted unit cost - which can then be weighed against other elements, such as environmental values. For comparison of similar projects, such as two direct supply reservoirs, the reliability criterion described above may be quite serviceable. But increasingly comparisons are being made between very dissimilar alternatives which will be incorporated in fairly complex systems of resources and will fail in different circumstances and perhaps with different consequences.

In long-term planning of resource strategy, such as that done by the former Water Resources Board, only the question of relative merit arises, since timing - within certain limits - is a matter for later determination.

OPERATING POLICY: MULTIPURPOSE OPERATION

Two aspects of operating policy have already been mentioned. These are: to what extent will the resources under one management authority be interlinked and used to give maximum output as a system, even at some increase in operating costs? and will there be an explicit restriction or rationing policy, to be applied when reserves have fallen to some pre-determined level and which can be incorporated in the planning and design of alternative projects?

One of the most important aspects of such a policy will be the allocation of dwindling resources between water supplies to consumers and residual river flows. At present water authorities generally plan on the premise that priority will be given to the maintenance of some minimum acceptable flow in rivers although it is well known that in practice priority will be given to the maintenance of the supply, using "drought" orders" to over-ride the provisions protecting the river. Nevertheless, factors other than the quantity of water available for supply will usually help to determine optimum use of a system, particularly by a multipurpose authority. These may include the quality of water in rivers, especially where there is a sequence of abstractions and effluent returns along the course of a river; the effects of reservoir level on amenity and recreation; the reservation of storage volume for flood control; and the variable flow requirements of fisheries, irrigation, hydro-power, etc. These can be incorporated in an explicit operating policy, after sub-optimisation, given that the advantages for these other interests are considered to outweigh the penalties for reliability of supply.

NEW CRITERIA OF RELIABILITY?

Some shortcomings of the criterion now in use have been mentioned:

(a) the simple model of failure takes no account of penalties in circumstances more severe than the design drought, or of the exigencies of operation during a drought of unpredictable duration;
(b) it is difficult to make a valid comparison of dissimilar projects or to evaluate a complex resource system in terms of the "probability of emptiness."

Furthermore the criterion takes no account of the additional flows available in most years. These can furnish additional supplies of lower reliability which may be quite appropriate for certain uses, such as irrigation. Fundamentally different projects may offer different facilities in this respect, in which case a single-valued yield/reliability criterion is inadequate to describe them.

A similar but separate point arises because demand is rarely stationary and resources are developed in steps to meet a normally smoothly growing demand. Since large steps produce a greater and more prolonged excess of resources before design demand is reached, a succession of large projects must give a lower average risk of failure, other things being equal, than a succession of smaller ones. The supply taken from Vyrnwy Reservoir - an extreme case - did not reach design rate until about half a century after commencement of operation and during the first few decades the risk of emptiness must have been infinitesimal.

Criteria have been suggested which would accommodate some of these difficulties; although it seems likely that the more completely they do so the less they lend themselves to the application of statistical theory. They may be based on the accumulated period of reservoir drawdown below a level representing the threshold of restriction, as a percentage of time (Jamieson, Radford and Sexton, 1964) or the accumulated shortfall of supplies, as a percentage of requirements, over a long test period. The "shortage index", for instance, is a function of the accumulated squares of supply deficits over 100 years (publications of US Corps of Engineers). The shortage criterion appears to have the advantage over the period criterion that it both provides a more complete reflection of impact of the failure and is likely to be less affected by the choice of restriction policy, since all active policies are likely to aim at minimising shortage but none (by definition) at minimising the period of restriction.

If simulation was to incorporate specific operating policies and if shortage criteria were elaborated to include penalties other than water supply deficits we might arrive at a composite index of performance - or its converse, an "index of deprivation" - which would be reasonably representative of all benefits/penalties. In effect the reliability problem would be subsumed in a general cost/benefit comparison of alternative projects and operating policies. It would depend on subjective trade-offs between one penalty and another and would be subject to the other well-known pitfalls of cost/benefit analysis. Nevertheless it might provide a more practical and versatile tool than "frequency of failure".

There might be little left for the statistician in this, except perhaps for the generation of suitable test series, with appropriate cross-correlation of data at different sites.

A more practical objection arises from the circumstances of the public local inquiry, where there is already sufficient difficulty in conveying the simpler notion of an average recurrence period of failure.

PRACTICAL APPLICATION: THE PUBLIC INQUIRY

It is at the public inquiry, or its equivalent elsewhere, that proposals for resource development are put on trial, and arguments for and against should be presented as plainly and succinctly as possible. Theoretical analysis which does not influence the verdict there may be of little practical effect. A very cursory account of some proceedings at a typical inquiry may therefore help to put things into perspective.

The latest of a series of inquiries into a proposed pumped storage regulating reservoir at Carsington, in Derbyshire, took place recently. The reservoir would form part of a complex of resources based on tributaries of the River Trent (mainly the rivers Derwent and Dove) and serving several urban and rural distribution networks, most of which have now been vested in the Severn-Trent Water Authority. Two features of the proceedings have some relevance to this discussion. The first is the basis on which yields were assigned to existing resources and to the proposed reservoir; the second the alternative solutions which were canvassed and whose merits had somehow to be compared with those of Carsington Reservoir.

The existing resources comprise a group of three direct supply impounding reservoirs, with later additional catchwaters (i.e. intakes in adjoining catchments) in the upper reaches of the River Derwent; a group of three pumped storage reservoirs filled from the River Dove; and various smaller reservoirs, river intakes and underground resources. They are interconnected in ways which (with some important limitations) permit them to be operated as a system. The proposed reservoir could be filled from the Dove or the Derwent, or both, and could make regulating releases to the Dove or the Derwent, or both. The main alternatives proposed by objectors to Carsington Reservoir were the abstraction of the remaining dry weather flow of one or both rivers, perhaps aided by partial conversion of the Derwent direct supply reservoirs to river regulation; **additional** impoundment for river regulation in the upper Derwent; and the exploitation of underground storage by intermittent pumping from boreholes in the Nottinghamshire sandstones to regulate the Derwent, perhaps to be preceded by temporarily increased abstractions or by shared use of a reservoir (Empingham) recently completed by the neighbouring Anglian Water Authority.

The yields of the existing Derwent Reservoirs had been evaluated on the basis of a flow record dating from 1905, and specifically on the 18-month drought of 1933-34, which was thought to have a return period of about 100 years. The yield so calculated was subsequently increased to about 63 million gallons a day (mgd) to take account of the supplementary catchwaters and subsequently by about another 1.3 mgd to reduce the nominal return period of failure to about 50 years. Of the total, 16.66 mgd is allocated as compensation water, but this has been reduced by about a half on occasions when a shortage of water for supply seemed imminent. However, more recently simulation of the 1959 flows (a severe single season drought) had revealed that the extra yield attributed to the catchwaters would not be realised in such a year and that the combined inflow and storage would only sustain a gross yield of some 60.7 mgd, the critical period of drawdown being about 9 months. It was on this basis that a yield of 44 mgd to supply - 5.6 mgd less than the previously accepted figure - was assigned at the recent inquiry.

The 1959 flows have also been used to evaluate the resources of the Dove, where there is only a short record, but in this case it was considered necessary to reconstruct the record because of certain eccentricities in the charts. Some uprating of the Dove reservoirs resulted, counterbalancing the reduction of yields on the Derwent. The yield claimed for the proposed reservoir at Carsington was likewise based on the 1959 flows in the two rivers. At a previous stage of these inquiries, however, a different approach had been used to evaluate the yield of Carsington. This involved the assembly of a synthetic drought by taking individual dry months from various years of record of produce a test condition more severe than any on record, but which, in the opinion of the designers, was equivalent to a 1/50 event. This procedure has been used in the design of a number of projects in the past; but it involves a special hazard when simultaneous flows at two or more points in a river system are synthesised independently. On this occasion two such "records" were misused to make a case for one of the alternatives proposed to Carsington, the conclusion of which was vitiated by a serious mismatch of flow data during a few critical weeks.

Stochastic Storage: Management

It is of some interest that the first notable application of statistical theory to the evaluation of reservoir yield in Britain was made by R. W. S. Thompson, who was for many years Engineer to the Derwent Valley Water Board. He studied the probability distribution of runoff totals from the Derwent Reservoir catchments for overlapping periods of between 14 and 20 calendar months throughout the period 1905-1948 and by plotting these on a Hazen diagram concluded that with mean recurrence intervals of 50, 100 and 200 years between failures, yields would be (equivalent to) about 59.5, 57.5 and 55.5 mgd, respectively, with critical periods of about 17 months. He recommended the 1/200 year basis for design. In an alternative approach he arrived at a yield equivalent to about 58.5 mgd with a 1 in 10 chance of failure in any period of 100 years (Thompson, 1950).

The following, taken from promoter's evidence at the recent inquiry gives, I believe, a fair indication of where many planners and designers stand in present circumstances:

"Since reorganisation the Authority has been able to make the fullest simultaneous use of as many of its sources as are inter-connected. Their combined value is assessed by studying coincident flow records. The river flow data available in the Dove/Derwent region have been examined, re-processed and extended and a data bank produced for the years 1932-1975.

Computer simulations of the resource system using this bank indicate that different reservoirs would empty in different years. The condition encountered in 1959 would, however, prove to be the most severe over-all in this period and has been chosen as the design condition for yield calculations ......

Assessment of the reliability of such a regional system (that is to say, of how often the system as a whole would fail to meet a certain demand) is complex and techniques have not yet been evolved for a comprehensive solution. The effect of natural hydrological events on both surface and groundwater resources is constrained by reservoir capacities, pipeline sizes, pump capacities and prescribed flow conditions. It is considered, however, that the combined output will fall to that available in 1959 not more than once in every 30 to 50 years. A year such as 1959 accords reasonably well, therefore, with the normal basis of reservoir design."

What does this cautionary tale illustrate, apart from the gap between the more sophisticated theoretical approaches and what perforce happens in practice? One thing is the compelling power of the recorded drought experience. Whatever the conclusion of a statistical study it is difficult to put the recorded event aside. In practice, nearly all current yield assessments in this country are decisively influenced by a chosen piece of flow record, with or without some genuflection in the direction of a statistical study. And the more complex the circumstances the greater the reluctance to stray far from the familiar historic droughts. At the same time it is clear that changes in yield corresponding with a change from, say, 1/50 to 1/100 recurrence are liable to be overshadowed by other sources of uncertainty; in this case doubts about the accuracy of flow records, about the active storage available and about the selection of a 1/50 test sequence for the system all counted for more.

The Carsington case also illustrates the inadequacy of any single flow sequence for comparing dissimilar proposals which depend upon flows at various points in the system and over various critical periods. The period of accumulated flows which would be critical for performance of the system might vary from a few weeks (using dry weather flow) to 5 or 6 months (Carsington) or several years (underground storage), so that from this point of view no single drought is likely to be a valid test for them all. Moreover they would "fail" in different ways which might present different options by way of response. The proposal to abstract dry weather flows was to be made viable by substituting water of lower quality from the River Trent to meet industrial needs along the lower Derwent and its failure would then be manifested by cooling or corrosion problems at these plants. Carsington would fail by exhaustion of storage, as modified by restriction policy; the groundwater system probably by limitations of pumping capacity. The problem, then, is to bring these (and other variations not mentioned here) to some common denominator of risk, because there is little doubt that their cost ranking can be upset by meddling, within plausible limits, with risk criteria.

The above may seem to have strayed a long way from the mathematical approach to the stochastic storage problem. But I believe that the mathematician who wants to contribute to the solution of water management problems needs to know, as far as his time and patience will permit, how these problems impinge on the decision maker and how each meshes in with other areas of uncertainty and controversy. In other words, he must come into the kitchen. He may conclude that we are concentrating on the wrong problems. It is therefore as well to approach Professor Lloyd's paper with this background in mind.

USE OF STATISTICAL METHODS

Statistical theory made its appearance rather late in the field of the hydrologist and the water engineer. But in the hands of many an enthusiastic amateur convenient statistical properties have been attributed to a variety of data, including not only natural data (rainfall amounts etc.) but also quantities resulting from the application of engineering constraints to these; for instance, the amounts pumped into storage in accordance with some legal formula, or the resulting state of storage. Alternatively, it has sometimes been assumed that the probability distribution of the natural phenomenon which leads to a certain condition can be transferred to the condition itself.

Many of these procedures have been severely criticised by statisticians, as well as by engineers who favour alternative methods, because they ignore correlations between data (adjacent or over-lapping runoff totals, for instance), because they too readily attribute distributions to highly "processed" output data, or for other reasons. Nevertheless the engineer must have some way of evaluating system performance in terms of its impact on the user, if he is to answer the questions: "is it time something was done?" and "what is the best thing to do?". It does not have to be expressed as a probability (or average recurrence interval) of failure if some other criterion of performance will be more readily understood.

There are several recognised problems in modelling the distribution of the input data (weekly, monthly or seasonal totals of river flow). However, the approach discussed by Professor Lloyd is to strive for an acceptable model of input data (reservoir inflows) and then to convert these, by mathematically respectable processes, into storage or yield distribution. Against this is the increasing complexity of some of the systems involved and the consequent remoteness of the output distribution. Is there any likelihood that the necessary transitions can be made or does the simulation alternative sketched out above offer a more realistic approach for these systems?

In any case there may be merit in treating the simpler storage problems more rigorously. That depends, among other things, on what difference it will make to the outcome in the case, say, of a choice between two direct supply reservoirs or in assigning a yield to such a reservoir. In Britain the relevant inflow units will usually be the accumulated inflows over periods of between 5 and 18 months, these being the periods over which inflow plus water available from storage is minimal, per unit of time. It would be instructive to see comparative treatments of some typical situations. Would Professor Lloyd come to a substantially different conclusion from that reached by R. W. S. Thompson, for instance, on the inflow data set out in Thompson's paper?

CONCLUSION

The purpose of this conference was to bring together theoreticians and practitioners. Both want to see new thinking applied in order to provide society with better answers. It may well be, however, that a comparison between this paper and some of the others will underline once more how far apart we are and how different the languages we speak. In my view a great deal of theoretical analysis in this field over the last few decades has produced a corresponding amount of literature but has had remarkably little impact at the business end. The blame, if any, must be shared, but I hope that the mathematician, for his part, will not be averse to looking at the ways in which the engineer has contrived some sort of answer to his problems - distasteful as these may sometimes be - and suggesting how he would have tackled the situation.

REFERENCES

Jamieson, D.G., Radford, P.J., and Sexton, J.R.,(1974),The Hydrological Design of Water Resource Systems: Water Resources Board.

Publications of U.S. Corps of Engineers, (various)

Report of the Royal Commission on Water Supply : Minutes of Evidence (1869), HMSO.

Studies of the Reliability of Water Supplies: Central Water Planning Unit (March, 1976). Since published as Interim Report on Studies of the Reliability of Water Supplies, January, 1977.

Thompson, R.W.S., (1950), "Application of Statistical Methods in the Determination of the Yield of a Catchment from Run-off Data," Institution of Water Engineers.

STOCHASTIC STORAGE PROBLEMS

E. H. Lloyd

*(Department of Mathematics,
University of Lancaster)*

ABSTRACT

After providing a brief critical summary of the basis and present state of analytical (as opposed to simulatory) stochastic reservoir theory, as introduced by Moran for independent inflows and developed by Lloyd for Markovian inflows, the paper goes on to show how the theory can deal with more elaborate inflow processes, including in particular those which are functions of a multivariate Markov chain.

The ARMA (1, 1) process is shown to be representable as a bivariate one-step Markov chain and hence an acceptable inflow for the reservoir model. This is noteworthy since this process can show Hurst-like behaviour.

It is suggested that in attempting to develop the theory to a realistic state the highest priority should be given to work on seasonal Markovian approximations to actual inflow processes. The paper discusses briefly the problem of finding parsimonious models for the transition matrices associated with such processes and points out that a possible alternative to autoregression is provided by a seasonal version of Pegram's matrix : here the distribution vector $v(i,t)$ of the flow $X(i,t)$ in the ith season of the t-th year is defined recursively as $v(i,t) = P(i) \, v(i-1,t)$ where

$$P(i) = \beta(i) I + \{1 - \beta(i)\} u(i) \, 1',$$

where $\beta(i)$ and $u(i)$ may be expressed in terms of the stationary seasonal inflow distributions and their lag-1 correlations.

******

The first task a theoretical applied probabilist ought to tackle is, I suppose, to find out what the real problems are.

He can then try to formulate them mathematically, and then try to solve them. Finding out about the real problems involves putting them in some kind of hierarchy: thus, I assume it is desirable to gain a general understanding of the behaviour of an interconnected system of reservoirs and then to provide algorithms or theorems giving some indication of the sensitivity of the system as a whole to selected variations, so that some kind of system optimisation could be attempted. But it is presumably more urgent to aim in the first instance at a nearer target, namely the behaviour of a single reservoir.

At this restricted level the primary aim is to provide a model that explains how a single reservoir actually operates, that is to express the relationship between storage and inflow/outflow characteristics under stated assumptions, and then to apply this to such problems as the following:

(i) in designing a new reservoir:

    to estimate the yield distribution, the relative frequency of failure (suitably defined), etc., as a function of size (or location, etc.) under specified operating conditions (which will not necessarily remain unchanged for ever!)

(ii) in operating an existing reservoir:

    (a) to estimate the yield distribution etc., as above,

    (b) to make relatively short-term predictions of the consequences of particular antecedents.

The distinction I have in mind here is between

    the distribution of "steady state" response to given (or assumed or estimated) inflow and demand <u>distributions</u>, on the one hand,

and

    the probable short-term future behaviour in the light of known recent inflow and demand <u>values.</u>

Let us consider the state of the existing theory in the light of these aims.

On the one hand we have methods based on simulation, as in the classical mass-curve method, in which a historical

Stochastic Storage: Theory

record of cumulative flow is examined graphically for its behaviour when subjected to various withdrawal policies (and reservoir capacities, if the capacity is a variable design parameter).

This method is completely free of assumption regarding inflow structure but raises some problems in the interpretation of its results, since these tell us how the reservoir would have behaved under specified management policies had it existed in the historical period covered by the data; whereas we want to know how it will behave in the future.

If the record is long enough (which in practice it rarely is) we can at least compile an empirical statistical picture giving the frequencies with which specified patterns occur.

Modernised versions of the above would use a computer instead of graphs, and would replace the historical record by a much longer synthetic record which simulates actuality, constructed on the plausible assumptions of classical time-series analysis, e.g., that the model should reproduce the actual historical seasonal means, with the superimposition of computer-simulated random fluctuations which preserve the historical variances and covariances: or possibly a more sophisticated assumption designed to reproduce the Hurst effect. The greater length of "record" here than in the last paragraph allows for a greater number of responses and an apparently sounder basis for making predictions from them; but this is of course sensitive to the accuracy of the assumed "data"-generating procedure. For interesting examples of early attempts at statistical simulation in hydrology see Hazen (1914).

In the so called "analytical" stochastic reservoir theory the reservoir content $Z(t)$ at time t is regarded as a stochastic process whose behaviour is defined by the inflow stochastic process $X(t)$ and the associated release policy, which may depend on current and past values of $X(t)$ and $Z(t)$ and also on other deterministic or stochastic "demand" variables.

The fundamental work described here is given in Moran (1954, 1955, 1959) and Gani (1957). We may summarise the development of this theory as follows:

Moran's basic model

The climate is supposed to be such that the year divides

itself into two clearly separated parts, a so called "wet season" and a "dry season." Inflow occurs only during the wet season.

X_t is the inflow available during the tth year, measured to the nearest integral multiple of a chosen basic volume unit. If X_t is large enough, overflow may occur. Withdrawal occurs during the dry season only, a fixed number (w say) of volume units being taken (or the total content, if this is less than w). The inflows $\ldots, X_{t-1}, X_t, X_{t+1}, \ldots$ are assumed to be mutually independent random variables with a common distribution. The storage sequence $\{Z_t\}$, where Z_t denotes the storage at the beginning of the t-th year, is governed by a stochastic difference equation expressing the conservation of volume:

$$Z_{t+1} = \min(Z_t + X_t, c+w) - \min(Z_t + X_t, w) \quad Z_t = 0, 1, \ldots, c.$$

(The same equations are obtained if instead of clearly separated wet and dry seasons we assume inflow and outflow to be proceeding simultaneously, provided the rates are constant over the working interval.)

It follows from the difference equation that $\{Z_t\}$ is a stationary Markov chain in which the distribution vector ζ_t of the storage Z_t is

$$\zeta_t = Q^t \zeta_0$$

where Q is a transition matrix whose elements are determined solely by the distribution vector ζ_t of the inflow X_t, the withdrawal w, and the capacity c of the reservoir. For sufficiently large t, (e.g., t>30, say), this implies that ζ_t converges to a fixed distribution vector ζ determined by the linear algebraic simultaneous equations $\zeta = Q\zeta$. All the relevant properties of the reservoir (yield, spillage, etc.) follow from $\{Z_t\}$.

The theory is a satisfactory one in the sense that, if the inflow and outflow assumptions were acceptable, we should have the steady-state storage <u>distribution</u>, namely the vector ζ such that $\zeta = Q\zeta$, and also a simple forecasting equation

Stochastic Storage: Theory 77

for the storage distribution at r time units ahead of a particular observed storage value: if Z_t is observed to have the value z the distribution vector of Z_{t+r}, for r = 1,2,..., is the zth column of the matrix Q^r.

Developments of the basic Moran model

Straightforward developments allow the same principles to be applied to a reservoir in which the inflows, whilst remaining mutually independent, are allowed to have seasonally varying distributions, and the withdrawals are allowed to depend on the current storage levels, again with seasonal variations; the working interval being reduced from a year to a month or a week or any convenient unit.

The most important assumptions that must be made if the theory is to be applied are:

(i)   the inflow distribution is known for each season;

(ii) the inflow in any one interval is independent of that in any others;

(iii) the working interval is short enough to allow us to adopt without serious error one or the other of the inflow/outflow sequencing models described under the heading

(iv) we may ignore the continuous nature of the flows;

(v)   we may neglect evaporation, blow-off, seepage and other losses;

(vi) no attention need be paid to secular changes in reservoir capacity due to silt deposition, etc.

Of these the most significant, methodologically, is *(ii)*, i.e., the independence assumption. If the historical record justifies this assumption for the working intervals adopted, the other requirements indicated by the other assumptions are not difficult to supply. It is true that the inflow distributions cannot be "known" precisely but they can be estimated by standard well documented statistical procedures, and the sensitivity of the resulting predicted behaviour of $\{Z_t\}$ to errors of inflow estimation may be obtained without too much difficulty.

The error involved in *(iii)* and *(iv)*, that is, what are the consequences of ignoring the detailed fluctuations <u>within</u> a working interval, do not seem to have been examined, but this ought not to present much of a problem.

Finally, one may express confidence in the ability of the model to absorb modifications that will allow for the water losses and changes of capacity outlined in *(v)* and *(vi)*; the required work (as far as I know) has however not yet been done.

[One further possible extension, not hitherto mentioned, would be an examination of the water quality, the "age" of various water strata, etc. As far as I know no attempt has yet been made to develop the theory in this direction.]

A contradiction

The usual way of reducing the error introduced by applying discrete-time methods to a continuous-time situation in applied mathematics is to make the working interval smaller. Unfortunately in the case of river flows, the smaller the working interval the less acceptable is the independence assumption: annual flows may perhaps be treated as being independent but daily flows cannot.

Serially correlated inflows

In order to be able to work with acceptably short intervals the model must somehow be modified so as to allow for a serial correlation in the inflow sequence.

The structure of the difference equation is conformable with Markovian assumptions, and it seems natural therefore to attempt to represent the serial correlation structure of the inflows by a Markov chain. This approximation appeals not only because it fits in well with the structure of the model but also because a Markov chain is a sufficiently flexible concept to be compatible with a wide variety of inflow distributions and serial correlation patterns.

If then we retain all the features of the Moran model and the outlined assumptions, with the sole difference that we replace the assumption of independent inflows X_t by the assumption that the inflow sequence $\{X_t\}$ is a (not

Stochastic Storage: Theory

necessarily homogeneous) Markov chain, we shall have made a substantial step in the direction of realism.

The storage process $\{Z_t\}$ is then no longer Markovian; but the sequence of <u>pairs</u> $\{(Z_t, X_t)\}$ is a (bivariate) 1-step Markov chain. That is, we may define a "label variable" U_t by some such code as the following

Z_t	0	0	...	0	1	1	...	1	...	c	...	c
X_t	0	1	...	n	0	1	...	n	...	0	...	n
U_t	0	1	...	n	n+1	n+2	...	2n+1	...	c(n+1)	...	(c+1)(n+1),

and $\{U_t\}$ is a simple Markov chain with matrix M, say, such that the distribution vector q_t of U_t (*i.e.*, the joint distribution of Z_t and X_t) satisfies the recursion

$$q_{t+1} = Mq_t$$

and converges to the steady state vector q such that

$$q = Mq.$$

Here the elements of the transition matrix M are completely determined by the release policy and the transition matrix L of the inflowing Markov chain $\{X_t\}$.

Thus, provided the assumption of Markovian inflows is acceptable, we have as before a satisfactory specification of the steady-state storage distribution and of the forecast joint distribution r time units ahead of an observed pair (z,x).

Once this has been appreciated, the way is open for further elaborations of the inflow and release processes. Multivariate Markovian inflows can be accommodated; and the release policy may depend on past as well as current inflow and storage values.

As an example suppose the inflow X_t to be a function – the sum, say – of two mutually independent Markov chains: $X_t = U_t + V_t$; with withdrawals depending on current inflow and storage values only. In this case the triplet (Z_t, U_t, V_t) is a trivariate 1-step Markov chain with properties corresponding to those described above for a bivariate chain, so that both the steady-state storage distribution and a forecast based on observed values are satisfactorily provided in terms of the transition matrices of the inflow chains. (Simple Markovian inflows were introduced by Gaver and Miller (1962), Kaczmarek (1963)and Lloyd (1963). For the extension to more elaborate Markovian inflows see Anis and Lloyd (1972) and Lloyd (1971).

Incorporating a Hurst-like inflow process

Of the stochastic processes known to demonstrate the Hurst effect, one of the simplest is due to O'Connell, who has shown empirically that this property is possessed by the simple (1,1) ARMA process:

$$X_t = \beta X_{t-1} + U_t + \gamma U_{t-1}$$

($\{U_t\}$ being a sequence of mutually independent random variables) for suitable values of the parameters β and γ.

Whilst this process is not a Markov chain, the vector (X_t, U_t) is. Arguments similar to those used later in this paper then show that if this ARMA inflow is fed into our reservoir, the triplet (Z_t, X_t, U_t) is 1-step Markovian. Its behaviour may be analysed in terms of the joint distribution vector p_t of the three variables, with

$$p_t = M p_{t-1} ,$$

M being a transition matrix whose elements are determined by the release policy and by the bivariate inflow transition probabilities $P(X_t = x, U_t = u | X_{t-1} = x', U_{t-1} = u')$ which in turn are specified in terms of β and γ by the structural equation for the ARMA process $\{X_t\}$.

We have here, perhaps (at last!), a technique for bringing together the analytic and the simulatory, two of the principal fields of endeavour in stochastic reservoir theory. If the representation of the inflows by an ARMA process is acceptable, we have - at least - a specification of the steady-state storage distribution from the trivariate distribution ector p defined by p = Mp. (It must be admitted that forecasting on the basis of observed values is less straightforward than with earlier examples, since here the "U_t" variables are latent and not directly observable.)

Stochastic Storage: Theory

Other workers have claimed the preservation of Hurst-like behaviour by even simpler processes such as the sum of two suitably chosen lag-one autoregressions. With such a process a device similar to the above would again be available, with the added virtue of allowing simple forecasts to be made in terms of an appropriate column of the matrix. [For the sake of brevity I have taken certain liberties in the last few paragraphs, treating X_t as discrete-valued (as required in our reservoir model) and, at the same time, as continuous-valued (as assumed in the ARMA equation). For consistency one ought of course to use a discrete-valued approximation to the ARMA equation.]

Other linear processes with Hurst-like properties may be dealt with in a similar way.

(Hurst's discoveries are described in Anis and Lloyd (1975, 1976), Hurst (1951) and Hurst, Black and Simaika (1965). For the theoretical work based thereon see Klemes (1974), Mandelbrot (1965), Moran (1964), Boes and Salas-La Cruz (1973), O'Connell (1971).)

Immediate practical problems

The most urgent of these are, in my opinion, the formulation of adequate models for inflow processes.

In those cases in which Markov chain models suffice the chief problem is that of estimating the inflow transition matrices. This replaces the more elementary requirement of having to estimate the inflow <u>distribution</u>, which was the corresponding problem in the case of independent inflows. Our problem is that of formulating few-parameter models for transition matrices, or, equivalently, few-parameter models for bivariate discrete probability distributions.

The obvious line to take in the absence of anything better is to use a matrix derived from a discrete-valued approximation to the familiar continuous-valued autoregressive scheme, the original non-Normal data being appropriately transformed. Seasonal versions of this scheme are well known.

We exemplify briefly an alternative approach in terms of distribution vectors. Firstly, a non-seasonal version: let $\{X_t\}$ be a homogenous Markov chain with distribution vector u_t

and transition matrix (Pegram (1972)

$$P = \rho I + (1-\rho) u\, 1', \quad (0<\rho<1,\ u \geqslant 0,\ 1'u = 1).$$

(Here $1'$ denotes the row-vector $(1,1,\ldots,1)$, u a column-vector, and $u\,1'$ a square matrix in which each column is identical with the vector u.) This has the intuitive attraction that it lies "between" two extreme cases of a Markov chain, one (with matrix I) corresponding to a completely deterministic chain and the other (with matrix $u\,1'$) corresponding to a chain of completely independent random variables. An increase in the parameter ρ increases the influence of the deterministic component and decreases that of the random component. In other words ρ is a measure of correlation. In fact it is not difficult to show that ρ is the lag-1 correlation coefficient of $\{X_t\}$, and u the stationary distribution vector. (The lag-h correlation coefficient is $\mathrm{corr}(X_{t+h}, X_t) = \rho^h$, $h = 1, 2, \ldots$.)

The matrix P is expressed in terms of readily indentifiable parameters, namely the correlation coefficient ρ and the parameters involved in fitting the vector u to the observed distribution.

The model is more interesting in a seasonal context, say a two-season year in which the inflows in the two seasons of the t-th year are $X_t^{(1)}$, $X_t^{(2)}$, respectively, with distribution vectors $u_t^{(1)}$, $u_t^{(2)}$. We shall refer to season 1 as "winter," season 2 as "summer." We suppose the sequence

$$\ldots X_{t-1}^{(1)},\ X_{t-1}^{(2)},\ X_t^{(1)},\ X_t^{(2)},\ X_{t+1}^{(1)},\ \ldots$$

to be a non-homogeneous Markov chain in which

$$u_t^{(2)} = P_2 u_t^{(1)} \quad \text{and} \quad u_t^{(1)} = P_1 u_{t-1}^{(2)},\ t = 1, 2, \ldots$$

so that

    (a) the sequence $\{X_t^{(1)}\}$, $t = 1, 2, \ldots$ of winter inflows is a homogenous Markov chain with transition matrix $P_1 P_2$,

and

(b) the sequence $\{X_t^{(2)}\}$ of summer inflows is a homogenous Markov chain with matrix P_2P_1.

We may now take P_1 and P_2 to have the form

$$P_j = \beta_j I + (1-\beta_j)v_j 1', \quad (0<\beta_j<1, \; v_j \geqslant 0, \; 1'v_j=1), \quad j = 1,2.$$

Whilst β_1, β_2, v_1 and v_2 do not have quite such direct interpretations as do the corresponding quantities in the non-seasonal version, they are explicitly expressible in terms of identifiable distributions and correlations. In fact we find

$$P_1 = (\sigma_1 \rho_{12}/\sigma_2)(I - \xi_2 1') + \xi_1 1'$$

and

$$P_2 = (\sigma_2 \rho_{21}/\sigma_1)(I - \xi_1 1') + \xi_2 1',$$

where ξ_j is the stationary distribution vector in the jth season, with variance σ_j^2 ($j = 1,2$), whilst $\rho_{12} = \mathrm{corr}(X_{t+1}^{(1)}, X_t^{(2)})$ and $\rho_{21} = \mathrm{corr}(X_t^{(2)}, X_t^{(1)})$. Here $\delta_{12} = \delta_{21}$, but in the multi-season version with more than 2 seasons per year it is not in general the case that $\delta_{rs} = \delta_{sr}$.

(The model requires that the correlation between successive summer flows should have the same value as that between successive winter flows, this common value being $\rho_{12}\rho_{21}$.) The generalisation to a k-season year presents no difficulties.

Further work is needed to develop a repertoire of parametric transition matrix models, together with appropriate estimation procedures and an accompanying catalogue of the distributions and correlograms implied by these matrices, so that one could choose a suitable model just as one now chooses a univariate distribution family of appropriate shape from the well known families available.

To sum up, it is my view that the Moran model, extended as above to take account of inflow serial correlation by means of (possibly multivariate) Markov chains, with seasonally varying inflow and withdrawal patterns, is within striking distance of being applicable to practical design problems. The outstanding tasks required to bring practical application still nearer are

   (a) implied by the assumptions of the existing theory
       (listed earlier)

and

   (b) the extension of the catalogue of transition matrices.

The theory of single reservoirs will then be reasonably complete and the stage will be set for such tasks as the invention of optimal management policies for single reservoirs, and, more ambitiously, the generalisation of the theory to systems of interconnected reservoirs.

REFERENCES

Anis, A.A. and Lloyd, E.H. (1972),"Reservoirs with mixed Markovian-independent inflows", *SIAM J. Appl. Math.;* **22**, 68-76.

Anis, A.A. and Lloyd, E.H. (1975),"Skew inputs and the Hurst effect", *J. of Hydrology,* **26**, 39-53.

Anis, A.A. and Lloyd, E.H. (1976),"The expected value of the adjusted rescaled Hurst range of independent Normal summands", *Biometrika,* **63**, 111-116.

Boes, D.C. and Salas-La Cruz, J.D. (1973),"On the expected range and expected adjusted range of partial sums of exchangeable random variables", *J. Appl. Prob.,* **10**, 671-677.

Gani, J. (1957),"Problems in the probability theory of storage systems", *J. R. Statist. Soc.,* (B) **19**, 181-206.

Gaver, D.P., and Miller, R.G. (1962), "Limiting distributions for some storage problems", Studies in applied probability and management, eds. K.J. Arrow, S. Karlin and H. Scarf, Stanford University Press.

Hazen, A. (1914), "Storage to be provided in impounding reservoirs for municipal water supply", *Trans. Amer. Soc. Civ. Engrs.,* **77**, 1539-1669.

Hurst, H.E. (1951), "Long-term storage capacity of reservoirs", *Trans. Amer. Soc. Civ. Engrs.,* **116**, 770-799.

Hurst, H.E., Black, R.P., and Simaika, Y.M. (1965), "Long term storage", London, Constable.

Kaczmarek, Z. (1963) Foundations of reservoir management, (Podstarry gospodarki zbornikowegi). In polish, with French summary. *Archiwam Hydrotechnik,* **10**, 2-37.

Klemes, V. (1974), "The Hurst phenomenon: A puzzle?",*Water Resources Res.,* **10**, 674-688.

Lloyd, E.H. (1963), "Reservoirs with serially correlated inflows", *Technometrics,* **5**, 85-93.

Lloyd, E.H. (1963), "A probability theory of reservoirs with serially-correlated inputs", *J. of Hydrology,* **1**, 99-128.

Lloyd, E.H. (1971), "A note on the time-dependent and the stationary behaviour of a semi-infinite reservoir subject to a combination of Markovian inflows", *J.A.P.,* **8**, 708-715.

Mandelbrot, B. (1965), *"Une classe de processus stochastiques homothetiques a soi; application a la loi climatologique de H.E. Hurst",* C.R. Acad. Sc. Paris, **260**, 3274-3277.

Moran, P.A.P. (1954), "A probability theory of dams and storage systems", *Austr. J. of Appl. Sci.,* **5**, 116-124.

Moran, P.A.P. (1955), "A probability theory of dams and storage systems: modifications of the release rules", *Austr. J. of Appl. Sci.,* **6**, 117-130.

Moran, P.A.P. (1959), "The theory of storage", London, Methuen Co. Ltd.

Moran, P.A.P. (1964), "On the range of cumulative sums", *Ann. Inst. Maths., Tokyo,* **16**, 109-112.

O'Connell, P.E. (1971), "A simple stochastic modelling of Hurst's law", Int. Symp. on Math. Models in Hydrol. Warsaw.

Pegram, G.G.S. (1972), "Some applications of stochastic reservoir theory", Ph.D. Thesis, Univ. of Lancaster.

SPATIALLY DISTRIBUTED VARIABLES IN HYDROLOGIC MODELLING

J. Amorocho

(Faculty of Water Science and Civil Engineering, University of California, Davis)

INTRODUCTION

Two different motivations have guided the historical development of hydrology. In one, the goal is scientific enquiry; in the other, hydrology is used primarily as a tool for the attainment of practical goals in engineering and water resources planning. In the latter, one may be willing to sacrifice scientific rigour as long as the approximations used give rise to errors deemed tolerable in terms of their practical consequences on the ultimate goals. Here, the faithfulness of the representation of the hydrologic systems may be secondary; but reasonable representations of the physical functions involved still have considerable importance, since the assessment of hydrologic risks can only be made by reference to the model at hand. If the model lacks generality, the errors it may entail remain unknown when it operates outside the range of the historical data used in its calibration.

Natural catchments are complex systems involving mass and energy transfer across the land - atmosphere interface and across topographic and geological boundaries. Ideally, for a complete representation of such a system, it would be necessary to know its inputs (one of which is precipitation) for every point in time and space, the states of storage in each surface and ground element, and the motion of every fluid particle. To say that such knowledge has never been available from observation, and that the knowledge of the physics of the processes involved is incomplete, is to state the obvious. Consequently, in hydrology a certain amount of intuitive idealisation and simplification has always been necessary. An important simplification made frequently is the replacement of temporally and spatially distributed variables, as well as fields of energy and mass flow, with functions of time only. This "lumping" process underlies most hydrologic modelling today.

RAINFALL FIELDS

Experience has shown that the use of lumped variable models is relatively successful in some cases for the prediction of streamflow. The inputs are in the form of time series of rainfall; other hydrologic variables are measured at isolated points within the tributary catchment, and the output is the flow rate past a designated cross-section of the stream. The measure of success of the model is based on a comparison between the model output and historically observed flows past the same section. Experience has also shown that this success, loosely assessed as it may be, is not attainable in many other cases with lumped models. Among the reasons for these failures are the poor quality and quantity of the rainfall data, since a few rain gauge records may be insufficient to describe the action of the rain on the catchment. For example, the path of a storm may be such that it does not intersect the location of the rain gauge. If the gauge record is used as the input to the model, the latter will not predict a flood, whereas the storm, unobserved at the gauge location, may indeed have produced high flows.

Widely separated rain gauges have been for many years the only precipitation measurement instruments available. The comparatively recent advent of new methods of observation, notably a few, very dense, synchronised recording rain gauge networks, a few weather radar installations, and certain artificial satellites, has made it possible to gain access to images that give us a greatly improved image of the kinematics of storms. This information, combined with new advances in the knowledge of storm dynamics, give us hope for better descriptions of precipitation fields at ground level. What we have learnt so far indicates that these fields are complex, and that their generating processes contain random and deterministic components interacting in time and space in intricate ways. Studies of sequences of cyclonic and convective storms, based on data from the West Coast of the United States and from semi-arid regions in Arizona and New Mexico, have revealed some salient features of these systems.

In broad outline, it is known that in a large number of cyclonic storms, such as occur, for example, in North-Central California and the Eastern United States, precipitation occurs in a series of rapidly moving bands of high intensity, superimposed on a general low-intensity background. The bands are meso-systems roughly oriented parallel to the front lines, and are composed of individual cells or clusters of cells ranging from 2 to 5 miles in diameter, having typically a life cycle of 10 to 15 minutes. The motion of these bands is sometimes faster than the motion of the general storm system and the cells displace themselves in a direction approximately parallel to the bands. The above structural features of the storm are reflected by the precipitation patterns near ground level.

Clearly defined banded structures exhibiting prominent elongation in directions parallel to the frontal traces are not always present in frontal systems. In particular, scattered groups of isolated echoes are often observed in the weather radar scope, exhibiting much smaller sizes than the bands, which, by contrast, may have typical lengths of the order of 150 miles or more, and widths of 20 to 25 miles. It is conjectured that these smaller rain regions may have an internal structure similar to the bands. Hydrologically, however, they are not likely to elicit important runoff responses, except in small catchments which may lie in their path.

Each storm may be identified as a sequence of events associated with one cyclonic system. The time between storms is detected on the ground as the interval between two consecutive non-zero rain gauge sequences pertaining to identified cyclonic storms. The storm duration is the total width of rain sequences assignable to a single storm. This width may be composed of one or more bands which are identified as periods of continuous non-zero sequences, within the total storm.

Convective precipitation is the principal source of runoff from catchments in many arid regions as well as some non-arid tropical ones. Since the size of these storms may be considerably smaller than the areas of the catchments affected, the storm locations and their spatial distribution may become as significant in determining runoff sequences as the temporal distributions.

Analysis of convective storms in semi-arid regions of the US were performed by the author, based on data furnished by the Agricultural Research Service (ARS), Southwest Watershed Research Center, Tucson, Arizona. Digital readings of strip charts of the networks of recording rain gauges at Walnut Gulch and Alamogordo Creek were available for a number of storms. Special computer programs were prepared to convert these digital data, representing accumulated gauge catch at

irregular intervals, into precipitation intensity values at
5-minute intervals by multi-point interpolation and differentiation. From the computed 5-minute intensity values given
at synchronous times for each one of the rain gauges of the
networks, a special contouring computer program plotted instantaneous isohyetal maps for each storm, representing the
rainfall patterns at 5-minute intervals. Sequences of these
maps show that the spatial distribution of rainfall resembles
one or more well defined, slowly travelling cells, each of
which has a "cauliflower" structure. The cells evolve constantly over a life span of 20 to 100 minutes, characterised
by a fitful process of growth and decay, during which the
maximum rainfall intensities are always observed in the central
region.

Models to simulate sequences of individual storm events
and the corresponding storm fields over limited geographical
regions have been proposed (Amorocho and Slack, 1970; Amorocho
and Morgan, 1971; Amorocho and Wu, 1975; Amorocho and DeVries,
1975). In these models, the sequences of individual storm
events are simulated by Monte Carlo schemes using parameters
inferred from historical rain gauge records and synoptic
weather charts.

The interior structures of the cyclonic storms are simulated by random clusters of cells of simple geometries, which
grow, decay and are replaced according to prescribed rules.
These rules are inferred from weather radar and dense rain
gauge network data.

For convective storms, the general configuration of any
cell is represented by a surface having the properties of a
bivariate Gaussian distribution. This produces maps with
elliptical isohyets that form in effect smoothed images of
the original 5-minute storm maps. The parameters of the surface for each time frame are determined by maximum likelihood
fitting. Each parameter is taken as a random function of time,
whose form is inferred from the analysis of complete map sequences for as many cells as have been observed.

These models are now operational, but their use shows
that many questions remain unresolved, as discussed later.

CATCHMENT FLOW AND CATCHMENT STORAGE DISTRIBUTIONS

Comparatively recent studies give indications regarding the complex patterns that groundwater storage distributions may attain in natural catchments. These patterns vary seasonally and are revealed for example by the extent of emergence of the zone of full saturation at the ground surface. The rates of infiltration of rain waters into the ground depend on these patterns. Only non-infiltrated rainfall runs overland and eventually reaches the streams. During dry periods only the parts of the basin near the channels contribute to streamflow from underground storage; the rest of the catchment makes negligible contribution. The contributing region expands and contracts in a dendritic but somewhat irregular fashion, depending on the season and the previous precipitation patterns. The development of this concept (called the "variable source" or "partial area" concept), and its confirmation by field observations, has led to further re-evaluation of some of the older methods of modelling. One of the results of these findings is a better understanding of some of the discontinuities observed frequently in hydrologic systems behaviour. These discontinuities may now be explained better physically, but the problems of handling them mathematically and numerically have increased vastly. Perhaps the new theoretical developments in the treatment of structural stability and morphogenesis may lead to useful computational algorithms in this area. It is clear that although more is now known about the elements that should go into the implementation of more faithful catchment models, it is also certain that the complexity of such models may become very great indeed, and that the cost of their implementation will increase accordingly. The point is illustrated by a model (Amorocho et al., 1973) for use in semi-arid climate catchments, which utilises a convective storm model interfaced with a catchment model consisting of many linked land and stream elements. For a 2000-acre catchment in Arizona, the model contained 34 land segments and half as many stream reaches. Infiltration parameters, roughness coefficients, dimensions, slopes and other pertinent variables must be measured or estimated for each land segment and stream reach. The length of computer run and its cost for the complete simulation of a flood produced by a single storm cell are quite reasonable. However, in the process of matching recorded and computed hydrographs, it is necessary to modify several times the estimated parameters and variables to which the model is sensitive. Therefore repeated test runs are required because the model does not contain at present features permitting automatic optimisation. Clearly, the optimisation problem becomes increasingly difficult as the catchment is subdivided further and further in order to gain representational faithfulness.

The range of trade-offs between model complexity and model coarseness is narrowed by the natural behaviour of catchments. Hydrologic systems may be heavily damped. Thus, the fine detail of the input fields, as well as the effects of the inhomogeneity and variability of the basin features, are attenuated to various degrees. One factor in this attenuation is the relation between the catchment size and the dimensions of the storm features. Thus, for some basins (say in the range of 100 to 500 square miles), cyclonic storm bands, notwithstanding their clustered structure, appear to be grossly equivalent in the production of runoff to areally uniform rains with the same mean intensities. The same storms would be perceived as sequences of distinct cells with high spatial and temporal intensity gradients in small urban catchments. Accordingly, whereas lumped parameter models might suffice for design and operational purposes in the former case, they could lead to large underestimation of peak runoff production in the latter. In the United States, Eagleson (1976) made a significant contribution to the establishment of criteria for the assessment of spatial and temporal damping in catchment models through the analysis of a simplified case involving static storm shapes and linear attenuation. The mathematical developments necessary for a more general solution of the problem capable of handling time-variable fields and non-linear wave propagation are yet to come.

CONCLUSIONS

The above remarks illustrate a few of the problems involved in hydrologic modelling. On the one hand, excessive model simplification and lumping may give poor performance. On the other, the advantages of the representational faithfulness in complex distributed models are offset by parameter uncertainties and computational cost.

A complete list of the problems in need of solutions in hydrologic modelling would be enormous. Four topics for inquiry, as suggested by the limited discussion presented here, are summarised as follows.

1. Probabilistic short-term predictions of precipitation fields over an area are needed for the operational control of systems with fine detail such as urban storm drainage networks with auxiliary storages. This requires refined storm models. New developments in the mathematical theory of random clusters would be helpful in this regard.

2. To permit intelligent choices between various models and various degrees of parameter and variable lumping, new developments in the theory of systems are needed, especially regarding non-linear wave propagation and frequency responses.

3. The treatment of three-dimensional flow problems (in porous media and in surface waters) needs improvement for applications in aquifers, estuaries and lakes. New mathematical developments in the finite element theory are needed for this purpose.

4. In lumped parameter as well as in distributed parameter models, the problems of system discontinuities are often difficult to handle. It is suggested that new applied mathematical concepts such as those of catastrophe theory could be particularly useful in this area.

Progress on this modest list of endeavours would bring very large benefits to the users of hydrology.

REFERENCES

Amorocho, J. and Slack, J., Simulation of Cyclonic Storm Fields for Hydrologic Modelling., AGU Annual Meeting Washington, D.C., 1970.

Amorocho, J. and Morgan, D., Convective Storm Field Simulation for Distributed Catchment Models., Proc. Intl. Symp. on Math. Models in Hydrol., IASH Warsaw, Poland, Vol. 2:4/15, July 1971, pp. 1-21.

Amorocho, J. and Wu, B., Mathematical Models for the Simulation of Cyclonic Storm Sequences and Precipitation Fields. Proc. Natl. Symp. on Precipitation Aanalysis for Hydrologic Modelling, AGU, Davis, California, June 1975.

Amorocho, J. and DeVries, J., A Convective Precitation Model for Distributed Catchment Simulation. Proc. Natl. Symp. on Precipitation Analysis for Hydrologic Modelling, AGU, Davis, California, June 1975.

Amorocho, J., De Vries, J.J., Morgan, D., Simulation of Runoff from Arid and Semi-arid Climate Watersheds. Vol. 1, Water Science and Engineering Paper No. 3002. Dept. of Water Science and Engineering, University of California, Davis, June 1973, pp. 1-110.

Eagleson, P.S., (1967) "Optimum Density of Rainfall Networks", *Water. Res. Research*, 3, No. 4 1021-1033.

SPATIAL PROCESSES: RECENT DEVELOPMENTS WITH APPLICATIONS TO HYDROLOGY

K. Ord and M. Rees

(Department of Statistics, University of Warwick)

ABSTRACT

Spatial processes, both continuous and point forms, may be described in terms of conditional expectations (or regression functions), covariance structure or by their spectra. The various formulations are contrasted and their relative advantages noted. Various approaches to rainfall measurement are then considered, using the underlying spatial processes to generate appropriate interpolation methods.

Spatial point processes are then reviewed briefly, in particular the centre-satellite and doubly stochastic Poisson schemes. Finally, the application of these processes to the modelling of thunderstorms is outlined.

1. INTRODUCTION

In contrast to the vast literature on temporal stochastic processes, the study of spatial processes has been relatively neglected until recently. "Standard" texts such as Matérn (1960) and Matheron (1971) have had only a restricted circulation and Bartlett's (1975) monograph is the first theoretical treatment of the subject to be freely available.

In this paper we are concerned primarily with modelling spatial processes rather than inference (see Cliff and Ord (1973)). Those interested in the analysis of hydrologic time series should consult Lawrance and Kottegoda (1977).

In section 2 we present a general review of spatial models and then give some applications of particular models to the measurement of rainfall in section 3. Spatial point processes are described in section 4 and their uses in the study of thunderstorms outlined in section 5.

2. SPATIAL VARIATION

A spatial process may be described in terms of:
a) the set of possible recording locations (sites);
b) the nature of the sites themselves; and
c) the associated random variables.

Besag (1974) lists the different varieties of spatial process with examples of each type; we describe the major alternatives briefly but only consider in detail those relevant to hydrological applications.

2.1 Set of possible sites

There may be a finite set of point locations, which may lie upon a regular lattice or be scattered in an irregular fashion. A regular lattice commonly occurs under experimental conditions as in studies of the growth and development of plants or trees. When such studies are carried out under non-experimental conditions the sites may be located irregularly.

In place of point locations we may consider spatial aggregates (or regions). For example, many economic variables relate to countries, states or census districts. The boundaries of such regions may or may not have an important bearing upon the structure of the spatial process. Again the number of possible sites remains finite.

Finally, we might consider a process which is defined for all points in space such as rainfall or height above sea level. The number of sites at which data are gathered will remain finite although the process is defined at all coordinate points, so that the set of all possible sites is a continuum.

It is evident that these alternatives are not hard and fast. A city may be regarded as a point location for migration studies, but as a set of regions in the study of journey to work patterns. If a set of point locations is under consideration then the properties of the process must be specified conditionally, given the set of sites. However, when the set of sites forms a continuum it is possible to specify properties (such as the covariance structure) for the whole space even though the data are recorded at a finite number of sites. We have laboured this distinction since it affects the choice of approach in later sections.

2.2 Nature of the sites

As mentioned in section 2.1, the sites may be either points or regions. Even with point sites we may wish to draw inferences for regions. For example rainfall is recorded at various points (gauges) but we use these data to describe rainfall patterns over a region. It is evident that we must specify an aggregation mechanism. So, if $Y(x)$ denotes the random variable Y at the point x, we may define the aggregate $Y(A)$ for region A as

$$Y(A) = \int_{x \in A} Y(x) \, dx \, . \qquad (2.1)$$

Unfortunately, it is difficult to proceed operationally using (2.1) when the regions are irregular (see Cliff and Ord, 1973, pp. 161-62).

2.3 The random variables

At each site x, the random variable may be discrete or continuous. Often, discrete variables relate to the absence or (multiple) presence of an "individual", so that there are only a finite number of sites for which a non-zero value is recorded. We refer to such schemes as <u>point processes</u> and return to a discussion of such processes in section 4. When the variates are continuous we may describe a process in either an autoregressive or a covariance framework as follows.

Covariance specification

For variate $Y(\cdot)$ located at (x, t) and at (x', t') we define

$$E[Y(x,t)] = \mu(x,t) \qquad (2.2)$$

$$\text{cov}[Y(x,t), Y(x',t')] = c(x,x',t,t') \, . \qquad (2.3)$$

If Y is a normal (Gaussian) process, the mean and covariance functions serve to specify the process completely. In general, we refer to a <u>wide</u> sense specification if only the mean and covariance are specified, while a <u>strict</u> sense version requires knowledge of the joint density functions.

Autoregressive specification

Given n sites $x_1, \ldots, x_n$ the autoregressive form is

$$\Delta_t[Y(x,t)-\mu(x,t)] = \sum_i \beta(x_i,t)[y(x_i,t)-\mu(x_i,t)] + \varepsilon(x,t) \quad (2.4)$$

where Δ_t represents a difference or differential operator with respect to time and $\varepsilon(x,t)$ represents random noise. Typically, we assume that

$$E[\varepsilon(x,t)] = 0, \quad (2.5)$$

$\text{cov}[\varepsilon(x,t),\varepsilon(x',t')] = \sigma_\varepsilon^2$, if $x=x'$ and $t=t'$, 0, otherwise.
$$(2.6)$$

It is evident that the autoregressive scheme could be extended by the incorporation of higher order spatial and/or temporal differences (derivatives).

2.4 Stationarity

The specifications given in the previous section are very general and further restrictions need to be imposed before the models are operational. A convenient, and often plausible, simplification is to assume that it is the relative position in time and space which is important rather than the absolute position. This leads us to the concept of stationarity.

Wide sense stationarity

The process $Y(x,t)$ is said to be stationary in the wide sense if

$$E[Y(x,t)] = \mu$$

$\text{var}[Y(x,t)]$ is finite for all x and t, $\quad (2.7)$

and

$$\text{cov}[Y(x,t), Y(x',t')] = c(x-x', t-t'). \quad (2.8)$$

Often (2.8) will be further specialised to

$$c(x-x', t-t') = c_1(x-x')c_2(t-t'); \quad (2.9)$$

Developments in Spatial Processes

that is, the spatial covariance structure is invariant for any given time interval (and *vice versa*). For the autoregressive form (2.4), the effect of equation (2.7) is obvious, while equations (2.8) impose structural restrictions upon the $\{\beta(x,t)\}$.

Strict sense stationarity

If, for any n and for any set of sites $x_1,\ldots,x_n$ we require the joint density functions to satisfy

$$f[\,y(x_1,t),\ldots,y(x_n,t)\,] = f[\,y(x_1-d,t'),\ldots,y(x_n-d,t')\,] \quad (2.10)$$

for all d and for all t', then the process is stationary in the strict sense. For a Gaussian process it is evident that conditions (2.7) and (2.8) suffice to establish (2.10) but (2.10) is, generally, a stronger condition. In the remainder of this section, we consider only the wide sense conditions (though often we shall be considering Gaussian processes).

In the context of geological processes Matheron and his co-workers (c.f Matheron (1971) and the references given therein) argue that restriction (2.7) is too severe and should be relaxed to consider

$$Y(x,t) - Y(x',t) \quad (2.11)$$

as a stationary process. $Y(\cdot)$ is then said to be quasi-stationary. Matheron gives a description of the process in terms of the **variogram** (the covariance function for process (2.11)). Alternatively, we could introduce a state equation such as

$$\nabla_x^2 \,\mu(x,t) = \delta(x,t) \quad (2.12)$$

where ∇_x^2 represents an "appropriate" spatial differencing (or differential) operator, as in (2.13) below. A similar scheme has been used widely in time series analysis (c.f. Harrison and Stevens, 1976). The net effect of (2.11), or (2.12), is to produce a discontinuity at the (time or space) origin in the covariance function. Such local behaviour is not physically realisable (c.f., Brownian motion) but the distinction between such localised effects and variation on a larger scale is often worth making. The local linear representation implied by (2.12) represents a considerable weakening of the assumptions.

Finally, we note that it is sometimes appropriate to assume that a process is direction invariant (or isotropic). The covariance function (2.9) then reduces to

$$c_1(r) \; c_2(t-t')$$

where $r = \|x-x'\|^{\frac{1}{2}}$ represents the Euclidean distance between sites x and x'. More general distance measures may be incorporated, but this topic is not pursued here.

One principal reason for the introduction of stationarity (or isotropy) assumptions is to allow the parameters of the process to be estimated from the data. The application of such methods depends upon the investigator finding a parsimonious representation of the complexities of the real world process. Also, it allows use of the spectrum as a descriptor of spatial processes, as we now show.

2.5 Spectral representation

To show how a frequency representation may be employed, we restrict attention to a particular first order scheme. We define the spatial differencing operator as

$$\nabla_x^2 = \nabla_1^2 + \nabla_2^2 \qquad (2.13)$$

where

$$\nabla_1^2 Y(x_1, x_2) = Y(x_1+h_1, x_2) - 2Y(x_1, x_2) + Y(x_1-h_1, x_2) \quad (2.14)$$

and ∇_2^2 is similarly defined, for observations on a grid spaced (h_1, h_2) units apart. Thus, a first order Markov spatio-temporal process could be of the form

$$\Delta_t Y(x,t) + (\alpha^2 - \nabla_x^2) Y(x,t) = \varepsilon(x,t) \qquad (2.15)$$

where $\varepsilon(x,t)$ denotes white noise, as defined in equations (2.5) - (2.6). For theoretical developments, it is generally more convenient to operate in continuous space and time, so that (2.15) becomes

$$\{D_t + \alpha^2 - D_1^2 - D_2^2\} Y(x,t) = \varepsilon(x,t) \qquad (2.16)$$

where $D_t = \partial/\partial t$ and $D_i = \partial/\partial x_i$, $i=1,2$. Following Bartlett (1975, p29), if we let

$$H(w_t,w) = \sigma^{-1}\int \exp\{i(w_t + w_1 + w_2)\} Y(x,t) dx dt$$

$$= (\alpha^2 + iw_t + \tfrac{1}{2}w_1^2 + \tfrac{1}{2}w_2^2)^{-1}$$

then (2.15) yields the spectral density function

$$f(w_t,w) = \sigma^2 H(w_t,w) H^*(w_t,w) \qquad (2.17)$$

where H^* is the complex conjugate of H.

If $f(w_t,w)$ is integrated over t, we obtain the purely spatial spectrum as

$$(\alpha^2 + \tfrac{1}{2}w_1^2 + \tfrac{1}{2}w_2^2)^{-1}.$$

As might be expected both (2.17) and (2.18) display a slow decay from the origin, typical of first order processes.

If the underlying process is non-stationary, of form (2.11) or (2.12) say, then the spectrum would have a spike at the origin while differenced process would produce spectra like (2.17) and (2.18).

From the practical viewpoint, the spectral analysis of spatial processes requires a huge amount of data on a regular grid. This is rarely available and various methods of "gridding" irregularly spaced data have been developed; see, for example, Robinson (1975) and Hsu (1975). It is our view that such methods should be used with extreme caution until more is known of their operational characteristics. However, in hydrological work, the data problems are not necessarily so acute and the use of radar for rainfall measurement (c.f., Harrold et al. 1974) would seem a source of suitable data in an area where spectral methods are potentially useful.

3. SOME SPATIAL MODELS AND THEIR APPLICATIONS

In this section we shall look at some particular models and examine their uses in hydrology. In general, we shall assume that the spatial covariance function is the object of primary interest and so most of the discussion relates to purely spatial models.

3.1 Purely spatial models

The spatio-temporal scheme given by equation (2.16) is a natural first order isotropic model, although directional dependence may readily be introduced. The purely spatial auto-covariance function is proportional to

$$2K_0(\alpha r), \quad \alpha > 0 \qquad (3.1)$$

where K_0 is the modified Bessel function of the second kind (Matérn, 1960; Whittle, 1962). Technically, the autocorrelation function is not properly defined in this case, because of local behaviour at the origin. More generally, when the spatial spectrum is of the form

$$(\tfrac{1}{2}w_1^2 + \tfrac{1}{2}w_2^2 + \alpha^2)^{-(s+1)}; \qquad (3.2)$$

the corresponding purely spatial scheme is, in two dimensions,

$$(\nabla_x^2 - \alpha^2)^{\frac{1}{2}(s+1)} Y(x) = \varepsilon(x) \qquad (3.3)$$

while the autocorrelation function, for $s>0$, is proportional to

$$(\alpha r)^s K_s(\alpha r). \qquad (3.4)$$

Thus, the most intuitively appealing spatio-temporal process has an awkward covariance function (as does the "natural" form of (3.3) with $s = 1$) while the negative exponential autocorrelation, $\rho(r) = \exp(-\alpha r)$ is given by $s = \tfrac{1}{2}$ with an unnatural underlying process.

The purely spatial model corresponding to spectrum (2.18) is best represented by the conditional statements

$$E[\varepsilon(x) | Y(x'), \text{all } x' \neq x] = 0$$

where $\varepsilon(x)$ is white noise and

$$\varepsilon(x) = (D_1^2 + D_2^2 - \alpha^2) Y(x). \qquad (3.5)$$

Again, such a process is not physically realisable for x continuous, but it becomes operational whenever a set of distinct sites $(x_1, \ldots, x_n)$ is considered.

In one dimension it is straightforward to produce a discrete first order (lattice) model for $Y(x)$ in terms of $Y(x - h)$ and $Y(x + h)$. In two dimensions we would need to consider all points lying on a circle of radius h, centred at x. Clearly this is not feasible and usually we prefer to retain a linear model involving only the sites (immediately) adjacent to x. Such modifications affect, in turn, the form of the autocorrelation functions and, as a result, several schemes that approximate (3.5) have been considered, notably

$$E\{Y(x)|\} = \gamma[y(x_1-1,x_2)+y(x_1+1,x_2)+y(x_1,x_2-1)+y(x_1,x_2+1)] \quad (3.6)$$

where $E\{y|\}$ denotes the expectation of $Y(x)$ given all other values; see Bartlett (1971) and Besag (1972a). If $z_j = e^{iwj}$, j=1,2 model (3.6) yields the spectral density function

$$[1 - \gamma(z_1 + z_1^{-1} + z_2 + z_2^{-1})]^{-1}. \quad (3.7)$$

The correlation structure for this scheme is difficult to evaluate and Besag (1972b) has developed various approximate expressions based upon one sided processes. One simple alternative is to consider the scheme with spectrum

$$[1-\gamma(z_1+z_1^{-1}+z_2+z_2^{-1})+\gamma^2(z_1+z_1^{-1})(z_2+z_2^{-1})]^{-1}; \quad (3.8)$$

this scheme incorporates four further neighbours as shown in Fig. 1. The advantage of scheme (3.8) is that the spectrum factorises into $H(z)H^*(z)$ where

$$H(z) \propto [1-\alpha(z_1+z_2)+\alpha^2 z_1 z_2]^{-1} \quad (3.9)$$

and $\alpha = \gamma/(1+\gamma^2)$. From (3.9) we may obtain a one sided model, so that the autocorrelation function is readily found as

$$\rho(x_1,x_2) = \alpha^{|x_1|+|x_2|}, \quad (3.10)$$

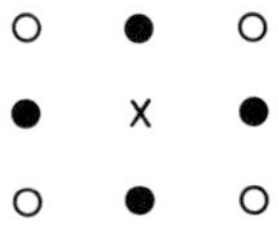

```
    O   ●   O
    ●   X   ●
    O   ●   O
```

Fig. 1 The nearest neighbour model (3.7) considers only the neighbours marked ●, whereas model (3.8) considers those marked O as well

involving a "city-block" distance metric. This and similar links between one and two sided models help to explain the better fits obtained for certain one sided schemes by Whittle (1954). See Whittle (1962) for further discussion of this point. Also, it could provide an easier approach to parameter estimation, although this remains to be investigated.

The other principal approach to spatial modelling, favoured by Matheron (1971), is to specify the form of $\rho(x)$ and then to derive conditional expectations given the particular set of sites at which observations were made. When the process is Gaussian, this will yield a regression function which is linear in the $\{Y(x')\}$ but non-linear in the parameters. Nevertheless, this approach is flexible and it is particularly suited to the analysis of data collected at an irregularly spaced system of sites.

3.2 Rainfall measurement

Eagleson (1967) used the autocorrelation function

$$\rho(r) = 1 + \beta_1 r + \beta_2 r^2 + \beta_3 r^3, |r| \leq r_0 \quad (3.11)$$

in his study of rainfall networks, following the approach of Matheron mentioned above. He then goes on to use spectral methods in a study of the optimal design of flood forecasting networks. As might be expected when isotropy is assumed, a regular grid of sites proves optimal, provided edge effects are ignored. In seeking an optimal density of sites, Eagleson appealed to the sampling theorem (c.f., Schwartz and Friedland, 1965, p.158) which states that we cannot detect wavelengths shorter than twice the grid spacing interval (the Nyquist frequency). Thus, Eagleson argues, we should decide upon the shortest wavelength requiring detection and lay out the grid accordingly. Given the slowly decaying nature of spatial spectrum (2.18), it seems unlikely that such a wavelength can be agreed very readily, so that the resulting decision may be subject to considerable uncertainty.

Mejia and Rodriguez-Iturbe (1974a) use autocorrelation function (3.4) with both $s = \frac{1}{2}$ and $s = 1$ in their study of the design of rainfall networks. They then proceed to develop estimators for long term areal mean rainfall. If we assume that the autocorrelation function in time and space factorises as in section 2.4 then we may consider the problem as one of spatial interpolation between rainfall aggregates for T time periods at each recording station. Given an isotropic process and a regular grid of n sites, the mean aggregate rainfall estimator

$$\frac{1}{nT} \sum_{j=1}^{T} \sum_{i=1}^{n} Y(x_i, t_j) \qquad (3.12)$$

will be very close to optimality. Also, given isotropy, a regular grid is to be preferred to random (stratified) sampling, as shown by Matérn (1960, p.83).

When the sites are spaced irregularly, a weighted estimator may be evaluated. This will be particularly important when only a few recording stations are available. Suppose that we wish to estimate the total rainfall over some region A, that is,

$$Y_A = \int_A Y(x) \, dx \qquad (3.13)$$

where the integral serves to aggregate and the time argument is dropped for convenience. We have available the observations at x_i, $i = 1, \ldots, n$ so we may specify the estimator

$$Y_A^* = \sum_{i=1}^{n} \lambda_i Y(x_i) \qquad (3.14)$$

which will be the best linear unbiased estimator (BLUE) if we minimise $E[(Y_A^* - Y_A)^2]$ subject to $\Sigma \lambda_i = 1$. When $E[Y(x)] = \mu$ for all x, we obtain the following estimators:

(a) for the overall mean, μ:

$$\hat{\mu} = \lambda^T y$$

where

$$\lambda^T = (\lambda_1, \lambda_2, \ldots, \lambda_n)$$

T denotes transposition and λ takes on the particular form

$$\lambda^T = (1^T V^{-1} 1)^{-1} 1^T V^{-1} \qquad (3.15)$$

where V is the covariance matrix for $\{Y(x_i)\}$ and

$$1^T = (1, \ldots, 1);$$

(b) for interpolation at any other point, x_0;

$$\hat{Y}(x_0) = \hat{\mu} + c^T V^{-1}(y - \hat{\mu}1) \qquad (3.16)$$

where $c = \{c(x_0, x_i)\}$, is the covariance vector, between $Y(x_0)$ and $\{Y(x_i)\}$;

(c) for total rainfall over some region A of area a:

$$\begin{aligned} Y_A^* &= \int_A \hat{Y}(x)\,dx \\ &= a\hat{\mu} + c_A^T V^{-1}(y - \mu 1) \end{aligned} \qquad (3.17)$$

where c_A has ith element $\int_A c(x, x_i)\,dx$.

When region A is regular and the covariances assume a simple functional form, it may be possible to evaluate c_A explicitly. Otherwise numerical integration is necessary. An example is presented in section 3.3.

While the estimators are BLUE if V is known, it will often happen that V (and c) will be functions of unknown parameters. In this situation, the estimators are no longer unbiased, but remain consistent.

This approach is termed Kriging by Matheron (1971, chapter 3) after the South African geologist D.G. Krige but is better known to statisticians as an extension of generalised least squares. This approach is more fully described in Matheron, who also extends the method (Matheron (1971) chapter 4) to cover spatial trends (universal kriging). A similar approach holds for quasi-stationary processes, provided that we seek to estimate differences rather than absolute levels.

This approach is broadly the same as that of Mejia and Rodriguez-Iturbe (1974b) who give several applications of the method, again using autocorrelation functions of form (3.4). Their conclusions suggest that the method is insensitive to the choice of autocorrelation function although the estimation methods used for the autocorrelation parameters were rather primitive.

Several other authors have discussed the design of rainfall networks from this viewpoint (Chemerenko, 1973; Huff, 1970; Osborn et al., 1972; Rumyantsev and Shanochkin, 1973) but their conclusions point to the need for a very dense network, particularly when convective storms are to be monitored. Huff (1970) stresses the value of radar measurements as an alternate data source.

3.3 An Example

Fig 2 shows a set of 8 irregularly placed sites around the River Wensum in Norfolk at which rainfall measurements are available. We are concerned in this example with two problems; that of interpolating rainfall values to other sites and that of estimating the total rainfall over a given area. For the former problem the estimation procedure (3.16) requires that we know V and c, the covariance matrix for the sample values and the covariance vector for the sample values and estimated point, respectively. The assumptions behind this procedure further require these covariances to be functions only of the distances between the respective points. If we assume the form for the autocorrelation known then point estimates are easily obtained. However, for the latter problem we must specify the covariance between the sample points and the over-all aggregate, namely c_A in (3.17), which has elements

$$\int_A c(x, x_i) \, dx = \int_A c(|x-x_i|) \, dx$$
$$= \sigma^2 \int_A \rho(|x_i - x|) \, dx$$

where $|\cdot|$ denotes the Euclidean distance. Thus we have the problem of integrating the autocorrelation function over the area A.

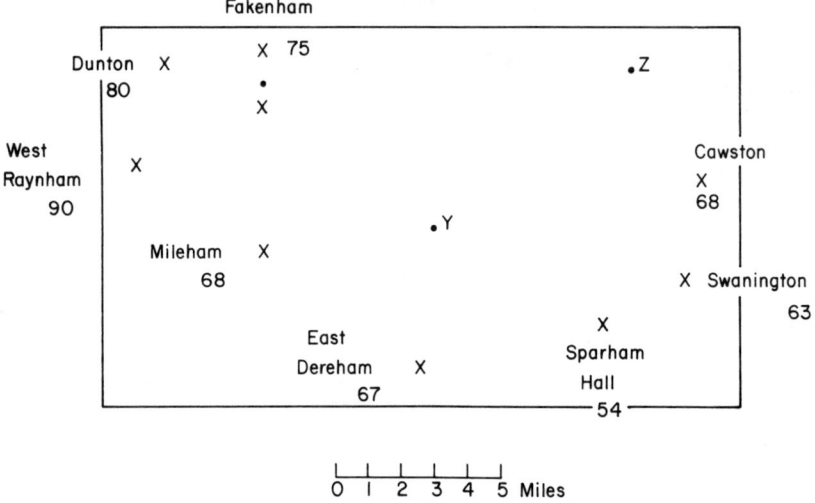

Fig. 2 Sites around the River Wensum with corresponding rainfall measurements (mm) for June 1968

To enable analytical results to be possible we have chosen A to be a rectangle enclosing the points. Many other boundaries are possible, for example the catchment area, but recourse would then have to be made to techniques of numerical integration to evaluate c_A.

Two forms of polynomial autocorrelation function were used in this example. This first is one of the simplest - the triangular autocorrelation function given by

$$\rho(|x-x_i|) = \begin{cases} 1-\gamma|x-x_i| & \text{for } |x-x_i| \leq 1/\gamma \\ 0 & \text{for } |x-x_i| > 1/\gamma \end{cases}$$

and the second is Eaglesons' cubic autocorrelation function (3.11).

The results are given in Table I for various parameter values. For convenience σ^2 was assumed to be unity throughout.

This example is not intended to be viewed as a case

Table I

Results of Estimation Procedures (3.15 - 3.17) Applied to the Data of Fig. 2

Autocorrelation Fuction	Interpolated Rainfall (Variance) mm			Mean Areal Rainfall (Variance) mm
	Site X	Site Y	Site Z	
Triangular ($\gamma=0.2$)	74.8(0.87)	68.7(0.17)	70.3(0.16)	69.4(0.17)
Triangular ($\gamma=0.1$)	74.9(0.93)	57.0(0.69)	67.3(0.52)	67.5(0.34)
$\rho(r)$ $=1-0.05r-0.006r^3$ ($|r|\leqslant 5$)	73.0(1.03)	65.8(0.23)	71.0(0.14)	70.1(0.14)

study, since the autocorrelation function would normally be unknown (see Section 3.2), but as an indication of the generalised least squares estimation procedure.

3.4 Rainfall runoff processes

In many ways the determination of runoff is even more important than the meansurement of rainfall itself. The runoff process is almost certainly non-stationary with strong dependence upon type of ground cover, nature of the soil, rock-bed and so on. Dincer et al. (1970) and Zmiyeva and Subbotin (1973) have attempted regression analyses for runoff while Amorocho and Brandstetter (1971) have employed an extension of the unit hydrograph method. The second method looks promising but may involve difficult estimation procedures.

A variation on the regression schemes produced so far would be to relate runoff (R) to precipitation (P) by a simple model such as

$$E(R) = \alpha + \beta P$$

but then to all β itself to vary according to the nature of the terrain. This approach has been used in geographical

4. SPATIAL POINT PROCESSES

Here we shall examine the stochastic processes, introduced in Section 2.3, where the discrete random variable associated with the sites relates to the absence or presence of an individual - such processes are known as point processes. The set of all possible sites is assumed to be a continuum.

Consider a point $(x, y) \in \mathbb{R}^2$, whose position is given by the vector r, surrounded by an infinitesimal rectangular area A; the orientation of area to point being the same for all possible pairs (x,y). Denote the size of the area A by $dr \triangleq dxdy$ and let $dN(r)$ be the number of individuals in A. If we assume the process to be spatially homogeneous, with mean density λ, then

$$\Pr\{\text{one individual in A}\} = \lambda \, dr$$

$$\Pr\{\text{more than one individual in A}\} \sim o(dr).$$

Hence, we may reasonably assume that the variable $dN(r)$ can only take the values 1 or 0 (corresponding to presence or absence of an individual, respectively). Thus all moments of $dN(r)$ are equal, notably

$$E\{dN(r)\} = E\{[dN(r)]^2\}. \tag{4.1}$$

A realisation of such a process will consist of a set of points, $\{(x_1,y_1),\ldots,(x_k,y_k)\}$, say. Associated with this set of coordinates will be the corresponding vectors $r_1, r_2, \ldots, r_k$ and infinitesimal cells $A_1, A_2, \ldots, A_k$.

The evaluation of probability relations between the various $dN(r_i)$ are greatly simplified, because of (4.1), since the product moment $E\{dN(r_1)\ldots dN(r_k)\}$ is non-zero only if there is an individual present in each A_i (i=1, 2, ...,k). Another consequence of (4.1) is that the product moment above represents the joint probability of an individual in each A_i. We can therefore define functions of the form

$$f_k(r_1, r_2, \ldots, r_k) dr_1 \ldots dr_k = E\{dN(r_1)\ldots dN(r_k)\}$$

where $f_k(\cdot)$ is known as the product density of degree k.

Whence,
$$E\{dN(r)\} = f_1(r)dr = \lambda\, dr.$$

The covariance density, $\gamma(r_1 - r_2)$, is defined by

$$\gamma(r_1-r_2)dr_1\, dr_2 = \text{cov}\{dN(r_1), dN(r_2)\}$$
$$= E\{dN(r_1)\, dN(r_2)\} - \lambda^2\, dr_1\, dr_2.$$

If isotropy is also assumed the covariance between the numbers of individuals in A_1 and A_2 becomes a function of the distance between A_1 and A_2, v say, alone, that is

$$\text{cov}\{dN(r_1), dN(r_2)\} = \gamma(v)\, dr_1\, dr_2$$

for example, if the points are placed randomly $\gamma(v) \equiv 0$, $\forall v > 0$.

Knowledge of the covariance density, $\gamma(v)$, and mean density, λ, permits us to obtain expressions for the expectations and covariances of the numbers of individuals in finite regions of the plane. Following Matérn (1971) if Q_1 and Q_2 are two such regions with areas D_1 and D_2 then the following moment formulae exist for the numbers of individuals, N_1 and N_2 say, in Q_1 and Q_2, respectively:

$$E(N_i) = \lambda\, D_i \quad \text{for } i=1,2$$

$$\text{cov}(N_1, N_2) = \lambda\, D_{12} + D_1 D_2\, E\{\gamma(v_{12})\}$$

where D_{12} is the area of intersection of Q_1 and Q_2 and v_{12} is the distance between two random points - one in Q_1 and the other in Q_2.

These formulae follow from the preceding argument concerning infinitesimal areas on dividing Q_1 and Q_2 into small cells and proceeding to the limit. Alternative expressions may be obtained by using the spectral representation of the covariance density (Bartlett, 1975, pp. 9-11).

Given a complete description of the mechanism giving rise to the spatial pattern λ and $\gamma(v)$ can normally easily be obtained. However, the distribution of the $\{N_i\}$ is almost always very difficult to handle and hence the derivation of nearest-neighbour distances, for example, is often intractable.

4.1 Centre-satellite Processes

In the centre-satellite structure we consider the <u>centres</u> to be realisations of a two-dimensional Poisson process and associated with each centre is a number of <u>satellites</u> (Neyman and Scott, 1958; Warren, 1971). The number of satellites corresponding to a particular centre is a random variable following an arbitrary discrete distribution. It is further assumed that the satellites are distributed spatially around the centre in accordance with some bivariate distribution which is a function of the distance between centre and satellite, $f(x-u, y-v)$ say, where (x,y) and (u,v) are the coordinates of the satellite and centre, respectively. The coordinates of the centre and satellite are random variables, and independence of all random variables is assumed.

It follows that the probability that a particular satellite, from a centre at (u,v), is to be found in an area A is

$$P(u,v) = \iint_A f(x-u, y-v) dxdy. \qquad (4.2)$$

If $G(t)$ is the probability generating function (p.g.f) of the number of satellites in a cluster and λ the density of the cluster centres, then the p.g.f. of the number of satellites from a centre at (u,v) in A is

$$H(t) = \exp\{\iint_A \lambda [G\{1-(1-t)P(u,v)\}-1] dudv\}. \qquad (4.3)$$

If we assume that $P(u,v)$ is such that satellites are always located at the cluster centre, then

$$P(u,v) = \begin{cases} 1 & \text{if } (u,v) \in A \\ 0 & \text{otherwise} \end{cases}$$

and

$$H(t) = \exp\{\iint_A \lambda[G(t) - 1]dudv\} \quad (4.4)$$
$$= \exp\{\lambda A[G(t) - 1]\}.$$

Thus, as we would expect, $H(t)$ is the p.g.f. of a generalised Poisson distribution, where $G(t)$ is the p.g.f. of the generaliser. Hence all the generalised Poisson distributions can be derived as special cases of the general centre-satellite model. For example, when $G(t)$ is the p.g.f. of the logarithmic series distribution (LSD), that is,

$$G(t) = \log(1-\beta)/\log(1-\beta t)$$

it follows from (4.3) that

$$H(t) = \{(1-\beta)/(1-\beta t)\}^k \quad (4.5)$$

the p.g.f. of the negative binomial distribution. When the individuals are identifiable with a particular centre, the spatial component can be ignored and the negative binomial is often a realistic model. When, as in hydrological modelling, the spatial component is important, we may generate the process in three stages:
  (i) locate the centres (u,v) by a Poisson process;
  (ii) generate the numbers per cluster using the LSD;
  (iii) locate individuals by generating co-ordinates (x,y) from the bivariate distribution with density $f(x-u,y-v)$.

One problem with the centre-satellite process is that the probability generating functional (4.3) can be obtained by a quite separate mechanism, known as the doubly stochastic process (Bartlett, 1964; Matérn, 1971). For this scheme, the initial Poisson has mean $\Lambda(x,y)$ for all $(x,y) \in \mathbb{R}^2$, where $\Lambda(x,y)$ also describes a stochastic process. Thus, the doubly stochastic process exhibits heterogeneity rather than clustering. However, both schemes lead to (4.3) so that the mechanisms cannot be distinguished by any single spatial data record, however large. Nevertheless, discrimination may be possible when data are recorded for several different points in time. For applications in hydrology (see the next section) both mechanisms appear to have something to offer in different circumstances.

5. A MODEL FOR THUNDERSTORMS

When clusters in the centre-satellite process are allowed to disperse spatially and become non-identifiable, the analysis becomes somewhat intractable. However, the process is well suited to simulation studies of the type described by Amorocho and Morgan (1971). Therefore, we consider the application of such processes to the modelling of thunderstorms.

It is well known that thunderstorms exhibit certain periodicities, the most pronounced being the annual periodicity. However, it is arguable whether there exist any other periodicities of a significant effect (LeCam, 1961). We shall therefore consider thunderstorms over short periods of time (less than 1 month) so as to enable us to ignore any seasonal effects.

Thunderstorms may be classified into one of three main categories:
1) Air-mass thunderstorms - storms in this category persist randomly within uniform air-masses.
2) Line thunderstorms - such storms congregate in narrow bands in the direction of wind at low levels.
3) Frontal thunderstorms - frontal thunderstorms are found in cloud region associated with a front and may be slightly more scattered than line thunderstorms.

In the majority of cases a thunderstorm consists of a cluster of thunderclouds in which the life of an individual thundercloud is quite short, 2 hours or less. However, this clustering has the effect of increasing the life of the storm far in excess of that of an individual cloud.

There exist many factors influencing the formation of thunderstorms, those of paramount importance being surface heating and the topography of the terrain. Whilst a deterministic prediction of thunderstorms is impossible there exist certain relationships between thunderstorms and terrain; for example, there is evidence that in mountainous regions airmass thunderstorms tend to predominate along the direction of a mountain range irrespective of the direction of the wind. It would thus seem possible to isolate certain uniform areas called storm activity regions by Amorocho and Morgan (1971) which have the same mean number of thunderstorms over a prolonged period.

In the brief description above we have seen that thunderstorms can occur at random, regularly, or in clusters within a region, and that there is evidence of possible heterogeneity over the given region.

In the case of air-mass thunderstorms, a single stage Poisson process may be adequate, at least within a single air mass. For a model over a wider area incorporating several air masses we may wish to generate the number of such air masses by a Poisson process and then allow a Poisson process within each air mass. Our model then becomes a simplified version of the doubly stochastic process.

Line thunderstorms may be obtained either by placing the centres in a more deterministic fashion or by restricting the Poisson scheme to a narrow band determined by wind conditions. A frontal storm then appears as a natural outgrowth of the line model by slightly widening the band.

To extend these purely spatial processes into spatio-temporal schemes we must incorporate time either directly into the model or indirectly through windspeed (LeCam, 1961). Finally, rainfall amounts may easily be generated by specifying an appropriate probability model for the amount of rain produced by cloud, such as the lognormal distribution (LeCam, 1961).

6. CONCLUSIONS

In this paper it was not our intention to provide new solutions, but rather to review the state of the art and to indicate possible directions for future research. It is unlikely that the full complexities of our climate will be ever encapsulated into a single model capable of providing worthwhile predictions. Therefore we must proceed by a mixture of simplifying assumptions and theoretical improvements which will, in the words of Professor Russell Ackoff, enable our model "to duplicate the essence of a system without actually achieving reality itself". We feel that models of spatial processes will prove a useful ingredient in the parsimonious development of such models.

7. ACKNOWLEDGEMENTS

It is a pleasure to acknowledge our debt to Professor Jaime Amorocho of the University of California at Davis and to Mr. John Cole of the Water Research Centre at Medmenham, Berkshire, for their efforts in introducing us to the hydrological literature on spatial processes and current problems of interest. The deficiencies of the present paper are, of course, our responsibility alone.

8. REFERENCES

Amorocho, J. and Brandstetter, A. (1971)"Determination of non-linear functional response functions in rainfall - runoff processes", *Water Resources Research*, **7**, 1087-1101.

Amorocho, J. and Morgan, D (1971) "Convective Storm field simulation for distributed catchment models", *Proceedings of the International Symposium on Mathematical Models in Hydrology,* Vol. **2**, 4/15, 1-21.

Bartlett, M.S. (1964), "The spectral analysis of two dimensional point processes", *Biometrika,* **51**, 299-311.

Bartlett, M.S. (1971), "Physical nearest-neighbour models and non-linear time series", *J. Applied Probability,* **8**, 222-232.

Bartlett, M.S. (1975),"The Statistical Analysis of Spatial Pattern,"London: Chapman and Hall.

Besag, J.E. (1972a), "Nearest-neighbour systems and the autologistic model for binary data", *J. Royal Statistical Soc.,* **34**, series B, 75-83.

Besag, J.E. (1972b), "On the correlation structure of some two dimensional stationary processes", *Biometrika,* **59**, 43-48.

Besag, J.E. (1974), "Spatial interaction and the statistical analysis of lattice systems", *J. Royal Satistical Soc.,* **36**, series B, 192-236 (with discussion).

Chemerenko, Y.P. (1973), "Errors in the averaging of data on the water equivalent of snow over the area", *Soviet Hydrology (Selected Papers),* **3**, 236-242.

Cliff, A.D. and Ord, J.K. (1971) "A regression approach to univariate spatial forecasting" in "Regional Forecasting"(eds. M.D.I. Chisholm *et al.*) London: Butterworth, 47-70.

Cliff, A.D. and Ord, J.K. (1973)"Spatial Autocorrelation," London: Pion.

Dinçer, T., Payne, B.R., Florkowski, T., Martinec, J. and Tongiorgi, E. (1970), "Snowmelt runoff from measurements of

Tritium and Oxygen 18", *Water Resources Research,* **6**, 110-124.

Eagleson, P.S. (1967), "Optimum density of rainfall networks", *Water Resources Research,* **3**, 1021-1033.

Harrison, P.J. and Stevens, C.F. (1976), "Bayesian Forecasting", *J. Royal Statistical Soc.,* **38**, series B, 205-247, (with discussion).

Harrold, T.W., English, E.J. and Nicholass, C.A. (1974), "The accuracy of radar-derived rainfall measurements in hilly terrain", *Quarterly J. Royal Meteorological Soc.,* **100**, 331-350.

Hsu, M.L. (1975) "Filtering process in surface generalisation and isopleth mapping" in "Display and Analysis of Spatial Data," (eds.J.C. Davis and M.J. McCullagh), London: Wiley, 115-129.

Huff, F.A. (1970), "Spatial distribution of rainfall rates", *Water Resources Research,* **6**, 254-260.

Lawrance, A.J. and Kottegoda, N.T. (1977), Stochastic Modelling of riverflow time series", *J. Royal Statistical Soc.,* **140**, series A, 1-47 (with discussion).

LeCam, L. (1961), "A stochastic description of precipitation", *Proceedings, 4th Berkeley Symposium in Mathematical Statistics and Probability,* **3**, 165-186.

Matérn, B. (1960), "Spatial Variation", *Meddelanden fran Statens Skogsforskningsinstitut,* Band **49**, 1-144 (Reproduced as a separate volume, 1970).

Matérn, B. (1971), "Doubly stochastic Poisson processes in the plane", *Statistical Ecology,* **1**, (eds. G.P. Patil, E.C. Pielou and W.E. Waters), University Park, Pennsylvannia State University Press, 195-213.

Matheron, G. (1971), "The Theory of Regionalised Variables and Its Applications,"Fontainebleau, Centre de Morphologie Mathematique.

Mejia, J.M. and Rodriguez-Iturbe, I. (1974a), "The design of rainfall networks in time and space", *Water Resources Research,* **10**, 713-728.

Mejia, J.M. and Rodriguez-Iturbe, I. (1974b), "On the transformation of point rainfall to areal rainfall",*Water Resources Research,* **10**, 729-735.

Neyman, J. and Scott, E.L. (1958), "Statistical approach to problems of cosmology", *J. Royal Statistical Soc.*, **20**, series B, 1-43 (with discussion).

Osborn, H.B., Lane, L.J. and Hurdley, J.F. (1972), "Optimum gaging of thunderstorm rainfall in Southeastern Arizona", *Water Resources Research*, **8**, 259-265.

Robinson, J.E. (1975), "Frequency analysis, sampling and errors in spatial data" in "Display and Analysis of Spatial Data," (eds. J.C. Davis and M.J. McCullagh), London, Wiley, 78-95.

Rumyantsev, V.A. and Shanochkin, S.V. (1973) "Evaluation of the representativeness of the precipitation network in relation to the spatial interpolation of precipitation", *Soviet Hydrology (Selected Papers)*, **3**, 199-206.

Schwartz, R.J. and Friedland, B. (1965), "Linear Systems," New York, McGraw-Hill.

Warren, W.G. (1971), "The centre satellite concept as a basis for ecological sampling", *Statistical Ecology*, **2**, (eds. G.P. Patil, E.C. Pielou and W.E. Waters), University Park Pennsylvania State University Press, 87-118.

Whittle, P. (1954), "On stationary processes in the plane", *Biometrika*, **41**, 434-449.

Whittle, P. (1962), "Topographic correlation, power law covariance functions and diffusion", *Biometrika*, **49**, 305-314.

Zmiyeva, Y.S. and Subbotin, A.I. (1973), "Spatial irregularity of spring runoff and losses in a river basin", *Soviet Hydrology (Selected Papers)*, **3**, 214-235.

MULTIVARIATE SYNTHETIC HYDROLOGY:

A THEORETICAL VIEWPOINT

Robin T. Clarke

(Institute of Hydrology, Wallingford, Oxfordshire)

1. THE PROBLEM

A water resource system may be regarded as a network N in which the nodes are either (a) sources, at which the "input" is usually streamflow or a volume of water pumped from an aquifer; or (b) sinks, where water is abstracted from the system ("output") to meet demand at centres of consumption by industry, agriculture or households; or (c) storages with finite volume. The inputs at the source nodes will be random variables, and the demands at the sink nodes will be dependent upon the inputs insofar as shortages in the supply of water to the network will lead to restrictions on its use. Links joining the nodes of the network, representing routes for water transfer from sources of supply to centres of demand, will be unidirectional and possibly subject to inequality constraints - as where the rate of transfer may not exceed some maximum value - or time constraints, as where transfers may only be effected during times of favourable electricity tariff for pumping.

The water resources engineer responsible for the system will often require to calculate some measure of system performance represented by a quantity F; this may be the probability with which the system fails to meet a fixed demand at some time during a time interval of length T; or the expected magnitude of the deficit; or the expected time for which it becomes necessary to introduce standpipe operation. Besides deriving an estimate $\hat{F}$ of F, the engineer will also require some measure of the reliability of his estimate, such as var $\hat{F}$.

2. THE FORMAL SOLUTION

To illustrate points made later in this paper, we take a

particularly simple system consisting of two nodes with a streamflow input at one, and a constant demand $\tilde{X}$ at the other. The link joining the nodes is unconstrained, and we take as F the probability of failure to meet the demand $\tilde{X}$ at some time during the next N time periods (to avoid the burdensome use of the "time period", and to avoid cluttering the argument with considerations of seasonal variation in streamflow, we assume that the time period is a year, with $\tilde{X}$ the total annual demand). With x_t the annual streamflow that is available for use in the tth year, we than have

$$F \equiv \text{prob}(x_t < \tilde{X} \text{ at least once in a sequence } x_1, x_2, \ldots x_N).$$

Although this definition of F is that used in the subsequent development, we note in passing that the water resources engineer may need to calculate several measures of system performance; thus alternatives to F could be F', F", or F"', given by

$$F' \equiv \text{prob(longest run of } x_t \text{ values that are less than } \tilde{X} \text{ in a sequence } x_1, x_2, \ldots x_N \text{ exceeds R)};$$

or

$$F" \equiv 1 - \varepsilon(m)/N$$

where m is the number of periods out of N for which demand is not met;

or

$$F'" \equiv 1 - \varepsilon(\Sigma \Delta W_i)/TD$$

where ΔW_i is the deficit in the ith period out of N for which the total demand is TD, and $\varepsilon(.)$ denotes an expected value.

However F is defined, a formal solution to the problem of estimating F is always possible, although usually of little value for computational purposes. To illustrate, consider the above definition of F and let the N-variate probability density of $x_1, x_2, \ldots x_N$ be

$$p(x_1, x_2, \ldots x_N; \mu, \Sigma \ldots)$$

where μ is a vector of means, Σ is a matrix of variances and covariances, and where other parameters besides μ, Σ may be necessary to define the functional form of the density function. Defining the integral

$$\Pi = \int_{x_1 > \tilde{x}} \int_{x_2 > \tilde{x}} \cdots \cdots \int_{x_N > \tilde{x}} p(x_1, x_2, \ldots x_N; \mu, \Sigma \ldots) dx_1 dx_2 \ldots dx_N$$

then the value of F is given formally by

$$F = 1 - \Pi.$$

The evaluation of F by this formula requires the evaluation of an N-dimensional multiple integral; even for this simple case of a two-node network with a simple definition of F, therefore, the computational burden of its calculation becomes formidable. Statement of the formal solution to the problem of estimating F thus has little purpose, save for the fact that it illustrates the three steps common to all such problems, namely:

(i) specification of the N-dimensional probability density $p(x_1, x_2, \ldots x_N; \mu, \Sigma \ldots)$
 (i.e., <u>choose the model</u> describing the streamflow sequence);

(ii) estimation of the parameters μ, Σ ... in the selected model (i.e., <u>fit the model</u> using as data the available streamflow records);

(iii) calculation of the numerical value of Π, and hence F. (i.e., <u>use the fitted model</u> to compute the numerical value of the postulated measure of system performance).

Expressed in this way, it is clear that the structure of the problem is identical in many respects with that of other hydrological problems; for example, where it is required to estimate the flood of T-year return period from a sequence of M annual maximum instantaneous discharges (or "annual floods", assumed independently distributed) the first step is to choose a (univariate) probability density function $f(x,\theta)$ of annual floods; the second step is to estimate the parameters θ from the record $x_1, x_2, \ldots x_M$; and the third is to calculate F from the equation

$$I/T = \int_F^\infty f(x, \hat{\theta}) dx.$$

It was shown above that the formal solution to the problem of calculating $F \equiv \text{prob}(x_t < \tilde{X}$ at least once in a sequence $x_1, x_2 \ldots x_N)$ is equivalent to the numerical evaluation of a multi-dimensional integral. In fact, it is always true that whenever synthetic streamflow sequences are used to calculate a quantity F, a multi-dimensional integral is being evaluated numerically. At the risk of labouring the example still further, consider the case where N = 2, so that F is

$$F = 1 - \int_{x_1 > \tilde{X}}^\infty \int_{x_2 > \tilde{X}}^\infty p(x_1, x_2) dx_1, dx_2 .$$

The double integral may be represented graphically as the integration of $p(x_1, x_2)$ over the shaded area in Fig.1.

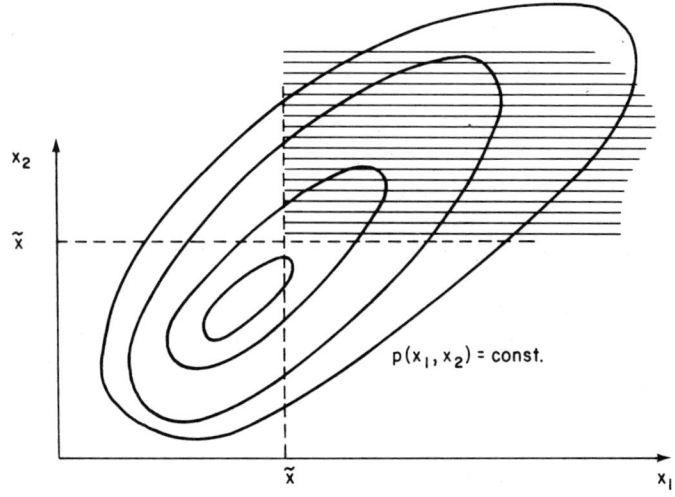

Fig. 1

Multivariate Synthetic Hydrology   123

With an integral of only two dimensions, it may be evaluated by a numerical integration rule such as Simpson's; for multi-dimensional integrals, however, it will usually be necessary to evaluate them by Monte Carlo methods. If the Monte Carlo method were to be applied to the trivial case of estimating the integral of $p(x_1,x_2)$ over the shaded area in Fig.1, one procedure (by no means the best) is to generate M synthetic sequences $(x_1^{(1)},x_2^{(1)}); (x_1^{(2)},x_2^{(2)});\ldots(x_1^{(M)},x_2^{(M)})$ and to record that number out of M for which either or both of x_1,x_2 were less than $\hat{x}$; if this number is m, then the (hit-or-miss) Monte Carlo estimate of the quantity F is

$$\hat{F} \equiv m/M .$$

2.1 Examples

(i) With $F \equiv \text{prob}(x_t < \tilde{x}$ at least once in a sequence $x_1,x_2,\ldots x_N)$, suppose that the x_t are generated by the lag-one autoregression

$$x_t - m = \rho(x_{t-1} - m) + \varepsilon_t$$

where the ε_t are Normally and independently distributed (NID) with zero mean and variance σ_e^2. Then the N-variate distribution $p(x_1,x_2,\ldots x_N; \mu,\Sigma\ldots)$ becomes multivariate normal with

$$\mu = [m,m,\ldots m]^T$$

$$\Sigma = \begin{bmatrix} \sigma_e^2/(1-\rho^2) & \rho\sigma_e^2/(1-\rho^2) & \rho^2\sigma_e^2/(1-\rho^2)\ldots \\ \rho\sigma_e^2/(1-\rho^2) & \sigma_e^2/(1-\rho^2) & \rho\sigma_e^2/(1-\rho^2)\ldots \\ \rho^2\sigma_e^2/(1-\rho^2) & \rho\sigma_e^2/(1-\rho^2) & \sigma_e^2/(1-\rho^2)\ldots \end{bmatrix}$$

....

(ii) With F as defined previously but with x_t generated by an ARIMA(1,0,1) model

$$(1 - \theta B)(x_t - m) = (1 + \phi B)\varepsilon_t$$

where ε_t is NID$(0, \sigma_e^2)$, then again p$(x_1, x_2, \ldots x_N; \mu, \Sigma \ldots)$ is multivariate normal with

$$\mu = [m, m, \ldots m]^T$$

$$\Sigma = \begin{bmatrix} \sigma_e^2(1+2\theta\phi+\phi^2)/(1-\theta^2) & (\theta+\phi)(1+\theta\phi)\phi_e^2/(1-\theta^2) \ldots \\ (\theta+\phi)(1+\theta\phi)\sigma_e^2/(1-\theta^2) & \sigma_e^2(1+2\theta\phi+\phi^2)/(1-\theta^2) \ldots \\ \ldots\ldots\ldots\ldots & \ldots\ldots\ldots\ldots \end{bmatrix}$$

Each model structure for the generation of values in the synthetic sequences $x_1, x_2 \ldots$ may therefore be regarded as defining a particular multivariate distribution p$(x_1, x_2, \ldots x_N; \mu, \Sigma \ldots)$.

3. MULTISITE MODELS

The subject of this paper is multisite streamflow generating models, but discussion so far has been restricted to the single-site case; however, it will be clear that extension to the multisite case involves only an increase in the dimensionality of the problem without a change in principle. In the multisite case, with sequences of N annual streamflows to be generated at each of K sites, the sequences

$$\begin{bmatrix} x_{11}, x_{12}, \ldots x_{1N}; \ x_{21}, x_{22}, \ldots x_{2N}; \ \ldots; \ x_{K1}, x_{K2}, \ldots x_{KN} \end{bmatrix}$$

can be thought of as one sample vector from a (K x N)-dimensional probability density function; as before, a model must be chosen which specifies $\mu, \Sigma \ldots$ and if the random variables ε in the model definition are NID$(0, \sigma_e^2)$ the vector of means μ and variance-covariance matrix Σ will determine the model completely.

Multisite streamflow generating models which specify μ, Σ are abundant in the hydrological literature. Probably the most widely used has been the multivariate lag-one autoregression

$$(x_t - m) + A(x_{t-1} - m) = B\varepsilon_t$$

where ε_t is a K-vector of independently and identically-distributed variates with zero mean, unit variance; x_t is a K-vector of synthetic streamflow values; and A, B are (KxK) square matrices (B triangular). Less widely used is the multivariate ARIMA model (O'Connell, 1974); the multivariate Thomas-Fiering model (Bernier, 1971); the multivariate fractional Gaussian noise model (Matalas and Wallis, 1971); the multivariate broken-line process (Mejia, Dawdy and Nordin, 1974); and the multivariate shot-noise model (Weiss, 1973). Because this paper is concerned more with principles than with particulars, individual characteristics of each model will not be discussed, but may be referred to in the papers quoted.

4. COMPONENT ERRORS IN THE ESTIMATE $\hat{F}$

When the quantity F is to be estimated as a measure of system performance, there will be at least three types of error to which the estimate $\hat{F}$ is subject. First, there will be errors because the model is at best an approximation to the more complex reality; second, because the model parameters θ must be estimated from data and so will be subject to sampling errors that would exist even if the model were a true description of reality; third, there will be errors arising from the Monte Carlo sampling procedure by which $\hat{F}$ is calculated, and these would be present even if the model and its parameters were determined with certainty. This paper is principally concerned with the second and third of these error components, but concludes with some tentative thoughts concerning the first.

4.1 *The effects on $\hat{F}$ of sampling errors in $\hat{\theta}$*

Adopting the classical approach, we first consider $p(\hat{F})$, the probability density of the estimate $\hat{F}$. We have

$$p(\hat{F}) = \int_{\theta \text{ space}} p(\hat{F}|\hat{\theta}) \, p(\hat{\theta}) d\hat{\theta}$$

$$= \int_{\hat{\theta} \text{ space}} p(\hat{F}|\hat{\theta}) \, p(\hat{\theta}; \theta) d\hat{\theta}$$

since the sampling distribution of the estimates $\hat{\theta}$ will depend upon the true parameter values θ. Hence we have

$$\varepsilon(\hat{F}) = \int_\theta p(\hat{\theta}; \theta) \, \varepsilon(\hat{F}|\hat{\theta}) \, d\hat{\theta}$$

and $\operatorname{var} \hat{F} = \int_\theta p(\hat{\theta}; \theta) \, \operatorname{var}(\hat{F}|\hat{\theta}) \, d\hat{\theta}$.

The mean value of $\hat{F}$ and its variance $\operatorname{var} \hat{F}$ are therefore weighted values of the conditional values $\varepsilon(\hat{F}|\hat{\theta})$ and $\operatorname{var}(\hat{F}|\hat{\theta})$.

If a Bayesian approach is adopted, with the posterior distribution $p_A(\theta|X)$ given data X and the prior distribution $p_B(\theta)$ related by

$$p_A(\theta|X) = (\text{constant}) \, L(X|\theta) p_B(\theta)$$

we have

$$p_A(F|X) = \int_\theta p_A(F, \theta|X) \, d\theta$$
$$= \int_\theta p_A(F|\theta, X) p_A(\theta|X) \, d\theta.$$

This yields the relations

$$\varepsilon(F|X) = \int_\theta p_A(\theta|X) \, \varepsilon(F|\theta, X) \, d\theta$$
$$\operatorname{var}(F|X) = \int_\theta p_A(\theta|X) \, \operatorname{var}(F|\theta, X) \, d\theta.$$

In discrete terms, the first of these becomes

$$\varepsilon(F|X) = \sum_{i=1}^{R} p_A(\theta_i|X) \cdot \varepsilon(F|\theta_i, X)$$

and the second

$$\operatorname{var}(F|X) = \sum_{i=1}^{R} p_A(\theta_i|X) \operatorname{var}(F|\theta_i, X).$$

The expected value and variance of the posterior distribution of F resulting from data X are therefore weighted sums of the expressions conditional on θ_i, a result analogous to

Multivariate Synthetic Hydrology

that given above for the non-Bayesian case. Clearly, any computational device leading to reduction in the conditional variances var $(F|\theta,X)$ will lead to a reduction in the value of var $(F|X)$ and later in this paper an account is given of two such devices tested on some trivial examples, which require further testing on real water resource systems.

Consider the problem of calculating a numerical value for $\varepsilon(F|X)$ by the above discrete relation, and suppose that the discrete approximation to the posterior distribution $p_A(\theta_i|X)$ is as illustrated in Fig. 2.

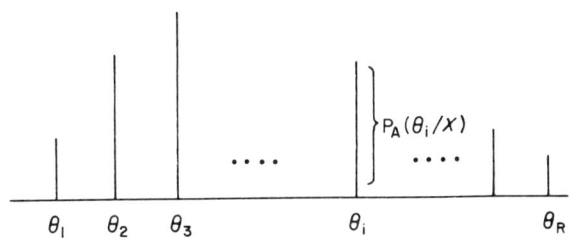

Fig. 2 Posterior probability density $P_A(\theta_i/\underline{X})$

For each θ_i (i = 1,2..R), the expected value $\varepsilon(F|\theta_i,X)$ must be estimated by Monte Carlo methods, and the estimates so obtained are weighted with $p_A(\theta_i|X)$ to obtain an estimate of $\varepsilon(F|X)$. A practical problem of some importance is therefore the following: given a sum of money sufficient for the generation of M* synthetic streamflow sequences: (i) how should R be chosen? (ii) how should the θ_i be chosen? (iii) how should the M* sequences be allocated amongst the R values θ_i for the calculation of $\varepsilon(F|\theta_i, X)$?

4.2 Errors arising from the Monte Carlo estimation of $\varepsilon(F|\theta_i,X)$

Two possible devices for reducing the conditional variance of F for given θ_i which have not been widely used in hydrological applications of Monte Carlo methods are the following:

(i) antithetic variates;

(ii) control variates.

Both devices belong to a class of "variance reduction methods" which have been described, for example, by Hammersley and Handscomb (1964). We present below some results following attempts to explore their potential for hydrological applications.

(i) Antithetic variates

Suppose that synthetic streamflow sequences are generated for input to a projected water resource system, and that the estimated measure of system performance is $\hat{F}$. The use of antithetic variates consists in seeking a second estimate of F, say $\hat{F}^*$, which is negatively correlated with $\hat{F}$; the two estimates $\hat{F}$ and $\hat{F}^*$ are then combined to form an estimate G, given by

$$\hat{G} = \tfrac{1}{2}(\hat{F} + \hat{F}^*).$$

If both $\hat{F}$ and $\hat{F}^*$ are unbiased estimates of F, then so is G, and if the negative correlation between $\hat{F}$ and $\hat{F}^*$ is large enough, var $\hat{G}$ will be less than var $\hat{F}$, the conventional Monte Carlo estimate.

A practical procedure for obtaining two negatively correlated estimates $\hat{F}$, $\hat{F}^*$ is illustrated by Fig.3. The generation of any pseudo-random variate by means of a computer subroutine has as its starting point the generation of a sequence of rectangularly-distributed pseudo-random numbers on the interval (0,1), say $\zeta_1, \zeta_2 \ldots$. The sequence is then transformed to (for example) Normally-distributed variates $u_1, u_2, \ldots$. Besides using the values $\zeta_1, \zeta_2 \ldots$ for transformation, the values of $1-\zeta_1$, $1-\zeta_2$, ... (which are perfectly negatively correlated with $\zeta_1, \zeta_2, \ldots$) may be similarly used, to give a sequence of Normally-distributed variates $u_1^*, u_2^*, \ldots$. The sequence $u_1, u_2, \ldots$ may be used to give the estimate $\hat{F}$, whilst the sequence u_1^*, u_2^*, ... may be used to give the estimate $\hat{F}^*$.

As a hypothetical example, consider the lag-one autoregression

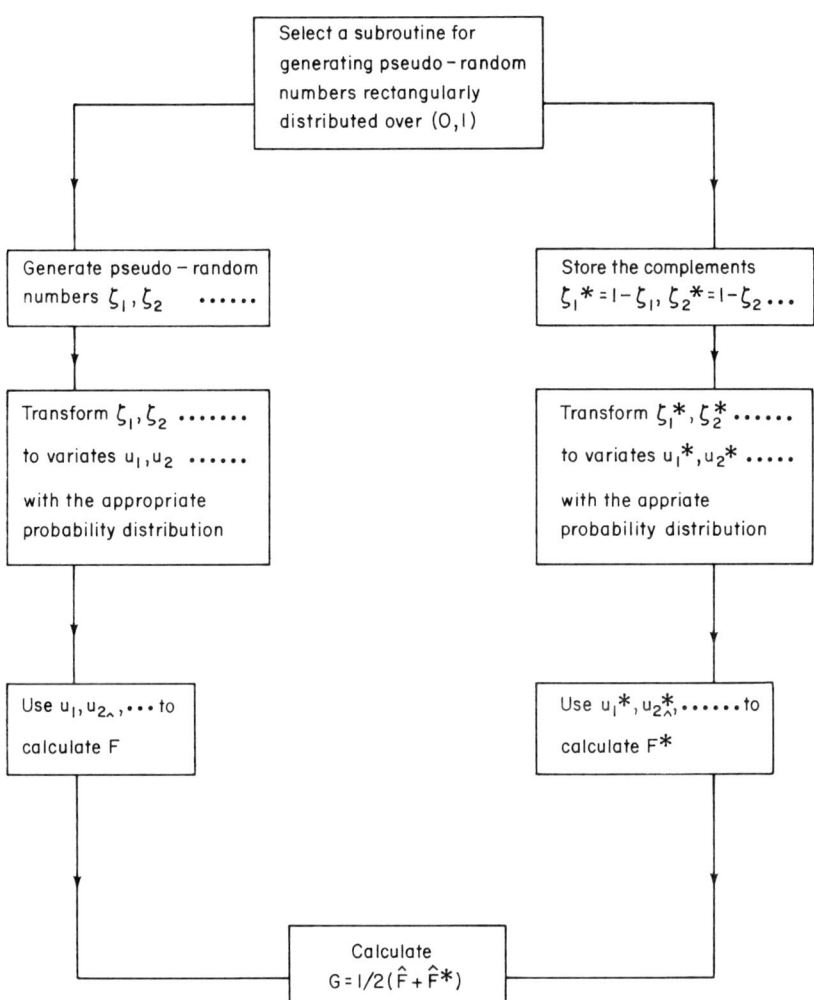

Fig. 3. Procedure for the derivation of negatively-correlated estimates F, F* for combination by the antithetic variate procedure

$$(x_t - 10) = \rho(x_{t-1} - 10) + \sqrt{9(1-\rho^2)}\, u_t \qquad u_t \sim N(0,1)$$

for which it is required to estimate

$F = \text{prob}(x_t > 17$ at least once in a sequence $x_1, x_2, \ldots x_{50})$.

TABLE I

Values of F, G obtained with the use of antithetic variates, and the same number of Gaussian variates generated for use in the model $(x_t - 10) = \rho(x_{t-1} - 10) + \sqrt{9(1-\rho^2)}\, u_t$ (For full explanation see text).

ρ	$\hat{F}$ (based on 5000 synthetic sequences)	$\hat{G}$ (based on 2 x 2500 synthetic sequences)	% reduction in variance
0.1	0.39 ± 0.0²691	0.39 ± 0.0²668	6.4
0.2	0.38 ± 0.0²688	0.38 ± 0.0²665	6.4
0.3	0.37 ± 0.0²684	0.37 ± 0.0²662	6.4
0.4	0.36 ± 0.0²678	0.35 ± 0.0²661	5.1
0.5	0.34 ± 0.0²668	0.34 ± 0.0²653	4.7
0.6	0.31 ± 0.0²655	0.31 ± 0.0²635	6.0
0.7	0.27 ± 0.0²631	0.27 ± 0.0²605	7.9
0.8	0.22 ± 0.0²588	0.21 ± 0.0²564	7.7
0.9	0.14 ± 0.0²486	0.13 ± 0.0²467	7.7

To calculate $\hat{F}$, 5000 sequences of 50 values $x_1, \ldots x_{50}$ were generated for each value of ρ from $\rho = 0.1(0.1)0.9$; to calculate $\hat{G}$, 2500 sequences were generated, and the variates 1-ζ were used to obtain an antithetic set; the same number of Gaussian variates were therefore generated as for $\hat{F}$. Table I shows the values of $\hat{F}$ and $\hat{G}$ obtained, their standard errors, and the percentage reduction in variance obtained (= 100(var $\hat{F}$ - var $\hat{G}$)/var F).

With the trivial example used, the percentage reduction in variance achieved by the use of the antithetic variate procedure is small; the cost of achieving this reduction in variance is, however, negligible, since a computer program for the calculation of $\hat{F}$ needs no more than a very slight modification to yield the antithetic estimate $\hat{G}$. If more practical simulation studies yielded similar variance reductions due to the use of antithetic sequences, then their routine application could perhaps be recommended, but the above trivial study would need much more supporting evidence before firm recommendations could be given.

(ii) Control variates

The control variate method consists of two steps. First, the estimation problem for the water resource system at hand is simplified to one that is capable of analytical solution; second, the analytical solution of the simplified system is used to improve the precision of the Monte Carlo estimate in the main system under study. As a practical procedure, the usual Monte Carlo estimate $\hat{F}$ is calculated by the generation of synthetic streamflow sequences; the same sequence of pseudo-random numbers generated for its calculation is used in deriving the estimate $\hat{F}^*$ of the simplified system, for which the expected value $\varepsilon(\hat{F}^*)$, denoted by ϕ, has been determined analytically. The control variate estimate is then $\hat{G}$, given by

$$\hat{G} = \hat{F} - \hat{F}^* + \phi.$$

The expected value of $\hat{G}$ is equal to that of $\hat{F}$, and provided that the estimates $\hat{F}$, $\hat{F}^*$ are sufficiently positively correlated, the variance of $\hat{G}$ will be smaller than that of $\hat{F}$.

As a hypothetical example, suppose that the fitted model is

$$x_t = \rho x_{t-1} + u_t$$

where u_t is NID(0,1); define F as

$$F \equiv \text{prob}(x_t > 2.3263 \text{ at least once in a sequence } x_1, x_2, \ldots x_{50}).$$

In the majority of sequences of total annual flow, the value of ρ is small, suggesting that a simplification of the fitted model would be that obtained by setting $\rho = 0$; the x_t of the simplified model then become

$$x_t = u_t$$

and for such a model (with u_t distributed N(0,1)) the value of ϕ can be written down immediately as

$$\phi = 1 - \left\{ \frac{1}{\sqrt{2x\pi}} \int_{-\infty}^{2.3263} e^{-\frac{1}{2}t^2} dt \right\}^{50}$$

which is readily evaluated from tables. Five thousand sequences were therefore generated for the calculation of the usual (hit-or-miss) Monte Carlo estimate $\hat{F}$ of F; the same pseudo-random numbers were used to calculate $\hat{F}*$ for the simplified model $x_t = u_t$, and the control variate estimate was obtained as

$$\hat{G} = \hat{F} - \hat{F}* + \phi.$$

TABLE II

Values of $\hat{F}$, $\hat{G}$ obtained with the use of control variates. The model used for F is $x_t = \rho x_{t-1} + u_t$ ($u_t \sim NID(0,1)$), and the control variate model is $x_t = u_t$, obtained by setting $\rho = 0$. (For full explanation, see text).

ρ	F (from 5000 sequences)	G (from 5000 sequences)	% reduction in variance
0.1	0.40 ± 0.0²624	0.40 ± 0.0²327	73
0.2	0.39 ± 0.0²644	0.38 ± 0.0²473	46
0.3	0.38 ± 0.0²656	0.38 ± 0.0²577	23
0.4	0.36 ± 0.0²700	0.36 ± 0.0²644	15
0.5	0.34 ± 0.0²727	0.34 ± 0.0²703	7
0.6	0.32 ± 0.0²699	0.32 ± 0.0²724	- 7

Table II shows the values of $\hat{F}$ and $\hat{G}$ obtained when ρ was given values $\rho = 0.1$ (0.1) 0.6, their standard errors, and the percentage reduction in variance resulting from the use of the control variate estimate.

Table II shows that considerable reductions in variance of the Monte Carlo estimate of F were obtained for small values of ρ, but that this reduction fell off rapidly as ρ increased to 0.5; this is to be expected, since the simplified model used to derive the control variate estimate had ρ set to zero. Alternative simplified models, giving alternative control variate estimates for which ρ was assumed small rather than zero, gave large reductions in variance of the Monte Carlo estimate which extended up to $\rho = 0.9$.

Multivariate Synthetic Hydrology

There appears, therefore, to be considerable scope for ingenuity in the construction of control variate estimates; and for the much-simplified estimation problems considered in this paper, the reductions in variance can be considerable. Whether such reductions could be obtained where the water resource system under study is of a real-life complexity, is a matter of some doubt, but one worthy of further study.

There is, however, a further very good reason for pursuing a study of the behaviour of control variate estimates to reduce the conditional variance of Monte Carlo estimates for given θ. This conference has heard a great deal in the past week of the enormous gulf separating the theorists on the one hand (who derive analytical results for reservoir systems that appear trivial by comparison with real-life systems) and the practising water resources engineer on the other, who works with such real-life systems on a regional basis. The essence of the control variate approach is that <u>analytical results from a simplified system are used to assist in solving the estimation problem for the much more complex system that is being simulated</u>; and it seems possible that the analytical results derived by the theorists could be used to assist the study of complex multi-storage systems.

5. THE PROBLEM OF MODEL CHOICE

This paper has discussed the effects of sampling errors in model parameters on the estimation of F, and also some methods for the reduction of the conditional variances of Monte Carlo estimates for given θ. There remains the difficult problem of model choice; if a set $M_1, M_2, \ldots M_R$ possible models is available for the description of the multisite streamflow inputs to a water resource system, what objective methods (if any) are available for the selection of a "best" model M_i?

There appears to me to be little alternative to approaching the problem of model choice through the avenues of statistical decision theory. My familiarity with its techniques is not great, so that the arguments given below are to be considered at best as a basis for much refinement, and at worst as needing refutation.

We assume that a decision D_j is to be selected from a set of possible decisions $D_1, D_2 \ldots D_P$. (For example, if P = 2, D_1 could be the decision to retain the existing operating rules for a water resource system, D_2 could be the decision to adopt a new proposed set of operating rules). We assume that the streamflow inputs to the system could be described by one of the set of models $M_1, M_2 \ldots M_R$; given historic streamflow records X, therefore, used to fit each model, a matrix of probabilities may be written of the form prob $(D_i^* | X, M_j)$, representing the probability that decision D_i is taken when model M_j is used after fitting to record X. This matrix is therefore as follows:

model:

	M_1:	M_2:	M_R:
D_1	prob$(D_1^*\|X,M_1)$	...	prob$(D_1^*\|X,M_R)$
D_2	prob$(D_2^*\|X,M_1)$	...	...
D_P	prob$(D_P^*\|X,M_1)$	...	...

A loss function must also be formulated which is a (PxP) matrix of values L_{ij}, where L_{ij} is the loss accruing where a decision D_j is taken and the optimal decision should have been decision D_i. This matrix is therefore the following:

decision taken:

		D_1:	D_2:	D_3:	...	D_P:
Optimal decision:	D_1:	0	L_{12}	L_{13}	...	
	D_2:	L_{21}	0	L_{23}	...	
	D_3:	L_{31}	L_{32}	0	...	
	D_P:	...	...			

Using the above two matrices, a (PxP) matrix of expected losses may be constructed, in which the (i,j)th element is the expected loss

Multivariate Synthetic Hydrology

$$\sum_{k=1}^{P} \text{prob}(D_k^* | X, M_j) L_{ik}$$

resulting from adopting decision D_i by the use of model M_j. This matrix is therefore:

	model: M_1:	M_2:	... M_R:
decision taken: D_1:	$\sum_k \text{prob}(D_k^* \| X, M_1) L_{1k}$	$\sum_k \text{prob}(D_k^* \| X, M_2) L_{1k}$	...
D_2:	$\sum_k \text{prob}(D_k^* \| X, M_1) L_{2k}$	$\sum_k \text{prob}(D_k^* \| X, M_2) L_{2k}$	...
.	.	.	.
.	.	.	.
D_P:			

If, on the basis of past experience, we can attach prior probabilities $\Pi_1, \Pi_2 \ldots \Pi_R$ to the models $M_1, M_2, \ldots M_R$ as measures of their suitability for the application at hand, then for each possible decision taken D_n we can form a risk given by

$$\sum_{i=1}^{R} \Pi_i \sum_k \text{prob}(D_k^* | X, M_i) L_{nk} \ .$$

That decision n should then be adopted for which D_n is a minimum.

Clearly the above argument, if valid, is at present of little more than academic interest because of the extreme computational burden of estimating the matrices of probabilities. Nevertheless, it appears that some such argument is necessary if the facts of water resource system simulation are to be incorporated; namely, (i) that the purpose of simulation is to select one decision from a series of possible decisions; (ii) that a loss (however difficult to quantify) will result if the wrong decision is taken; (iii) that a series of models for the simulation of streamflow input to the system is available, each of which could be fitted using the historic streamflow record X; (iv) that past experience may suggest preferences (or lack of preferences) for certain models of the set $M_1, M_2, \ldots M_R$, none of which is more than an approximation to the full complexity. Possibly some more heuristic approach is required in which long records of streamflow are divided into halves, with a set of models each fitted to the first half of the record; a comparison of the decisions reached using each model could be compared with that indicated by the use of the entire record.

6. FUTURE RESEARCH POSSIBILITIES

Topics for future research may be divided into those that are more mathematical, and those that are less so.

(i) It has been shown that where streamflow sequences are generated for the estimation of a measure F of system behaviour, the calculation is equivalent to the Monte Carlo evaluation of a multi-dimensional integral such as

$$\int_{x_1 > \tilde{x}} \cdots \int_{x_N > \tilde{x}} p(x_1, x_2, \ldots x_N) dx_1, dx_2 \ldots dx_N .$$

Theory is available by which a multivariable function such as $p(.)$ may be expressed in the form

$$p(x_1, \ldots x_N) = \sum_{i=1}^{R} g_{i1}(x_1) g_{i2}(x_2) \ldots g_{iN}(x_N)$$

and if such an expression could be found, the evaluation of the integral could perhaps be much simplified, since the multi-dimensional integral could be reduced to a sum of products of simple integrals. To what extent are such expansions of practical use in the simulation of water resource systems?

(ii) For a continuous Gaussian process $x(t)$, Cramér and Leadbetter (1967) give the values of $\varepsilon(Z_n(t))$, var $(Z_n(t))$ and the asymptotic distribution of $Z_n(t)$, where $Z_n(t)$ is given by

$$Z_n(t) = \frac{1}{T} \int_0^T \{x(t) - \tilde{x}\}^n dt.$$

Thus $TZ_1(t)$ is the shaded area in Fig.4.

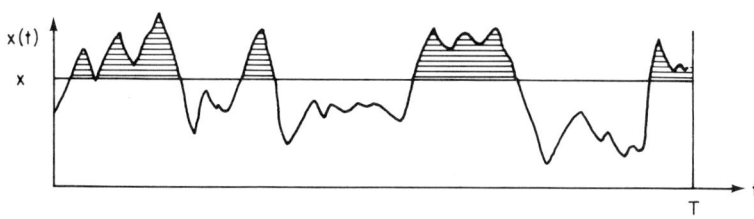

Fig. 4

However, the water resource engineer may well be concerned more with the distribution of the <u>maximum</u> excursion above (or below) the line defined by $\tilde{X}$ than with the properties of the <u>total</u> excursion above it. This distribution needs to be determined for Gaussian processes, and also for non-Gaussian processes.

Among the less mathematical topics:

(i) There appears to be much scope for the comparison of streamflow generating models fitted to a portion only of a long reliable streamflow record; the decision suggested by the use of each model could be compared with the decision suggested by the use of the whole record. If, for a particular estimation problem, one model consistently gave a good decision, there would be grounds for recommending this model for the particular estimation problem being considered.

(ii) For a given model with parameters θ, a given measure F of system performance, and a given number M of streamflow sequences that can be generated from a sum of money M* available for the estimation of F, there appears to be scope for study of a sampling method by which the M sequences should be allocated amongst the values θ_i of the prior distribution of θ, for the estimation of the conditional expectations. $\varepsilon(F|X,\theta_i)$ and conditional variances var $(F|X,\theta_i)$.

REFERENCES

Bernier, J. (1971), "*Modèles probabilistes à variables hydrologiques multiples et hydrologie synthétique*, IAHS Symposium Mathematical Models in Hydrology, Warsaw, July 1971, 333-343.

Cramér, H., and Leadbetter, M.R. (1967), "Stationary and Related Stochastic Processes: Sample Function Properties and Their Applications", John Wiley & Sons, Inc: London.

Hammersley, J.M., and Handscomb, D.C. (1964), "Monte Carlo Methods", Methuen: London.

Matalas, N.C., and Wallis, J.R. (1971), "Stochastic Properties of Multivariate Fractional Noise Processes", *Water Resour. Res* **7** (6), 1460-1468.

Mejia, J.M., Dawdy, D.R., and Nordin, C.F. (1964), "Streamflow Simulation 3: The Broken Line Process and Operational Hydrology", *Water Resour. Res.* **10** (2), 242-245.

O'Connell, P.E. (1974), "Stochastic Modelling of Long-term Persistence in Streamflow Sequences", Thesis submitted for the degree of Doctor of Philosophy - University of London.

Weiss, G. (1973), "Filtered Poisson Processes as Models for Daily Streamflow Data". Thesis submitted for the degree of Doctor of Philosophy - University of London.

Autoregressive specification

Given n sites $x_1, \ldots, x_n$ the autoregressive form is

$$\Delta_t [Y(x,t) - \mu(x,t)] = \sum_i \beta(x_i, t)[y(x_i,t) - \mu(x_i,t)] + \varepsilon(x,t) \quad (2.4)$$

where Δ_t represents a difference or differential operator with respect to time and $\varepsilon(x,t)$ represents random noise. Typically, we assume that

$$E[\varepsilon(x,t)] = 0, \quad (2.5)$$

$$\text{cov}[\varepsilon(x,t), \varepsilon(x',t')] = \sigma_\varepsilon^2, \text{ if } x=x' \text{ and } t=t', \; 0, \text{ otherwise.} \quad (2.6)$$

It is evident that the autoregressive scheme could be extended by the incorporation of higher order spatial and/or temporal differences (derivatives).

2.4 Stationarity

The specifications given in the previous section are very general and further restrictions need to be imposed before the models are operational. A convenient, and often plausible, simplification is to assume that it is the relative position in time and space which is important rather than the absolute position. This leads us to the concept of stationarity.

Wide sense stationarity

The process $Y(x,t)$ is said to be stationary in the wide sense if

$$E[Y(x,t)] = \mu$$
$$\text{var}[Y(x,t)] \text{ is finite for all } x \text{ and } t, \quad (2.7)$$

and

$$\text{cov}[Y(x,t), Y(x',t')] = c(x-x', t-t'). \quad (2.8)$$

Often (2.8) will be further specialised to

$$c(x-x', t-t') = c_1(x-x') c_2(t-t'); \quad (2.9)$$

sites. We have laboured this distinction since it affects the choice of approach in later sections.

2.2 Nature of the sites

As mentioned in section 2.1, the sites may be either points or regions. Even with point sites we may wish to draw inferences for regions. For example rainfall is recorded at various points (gauges) but we use these data to describe rainfall patterns over a region. It is evident that we must specify an aggregation mechanism. So, if $Y(x)$ denotes the random variable Y at the point x, we may define the aggregate $Y(A)$ for region A as

$$Y(A) = \int_{x \in A} Y(x) \, dx . \qquad (2.1)$$

Unfortunately, it is difficult to proceed operationally using (2.1) when the regions are irregular (see Cliff and Ord, 1973, pp. 161-62).

2.3 The random variables

At each site x, the random variable may be discrete or continuous. Often, discrete variables relate to the absence or (multiple) presence of an "individual", so that there are only a finite number of sites for which a non-zero value is recorded. We refer to such schemes as <u>point processes</u> and return to a discussion of such processes in section 4. When the variates are continuous we may describe a process in either an autoregressive or a covariance framework as follows.

<u>Covariance specification</u>

For variate $Y(\cdot)$ located at (x, t) and at (x', t') we define

$$E[Y(x,t)] = \mu(x,t) \qquad (2.2)$$

$$\text{cov}[Y(x,t), Y(x',t')] = c(x,x',t,t'). \qquad (2.3)$$

If Y is a normal (Gaussian) process, the mean and covariance functions serve to specify the process completely. In general, we refer to a <u>wide</u> sense specification if only the mean and covariance are specified, while a <u>strict</u> sense version requires knowledge of the joint density functions.